Traffic Engineering Design

Traffic Engineering Design

Principles and Practice

Second edition

Mike Slinn
MVA Limited, MVA House, Victoria Way, Woking GU21 1DD, UK

Paul Matthews
MVA Limited, Third Floor, One Berners Street, London W1T 3LA, UK

Peter Guest
8 The Grove, Farnborough, Hampshire GU14 6QR, UK

ELSEVIER
BUTTERWORTH
HEINEMANN

AMSTERDAM • BOSTON • HEIDELBERG • LONDON • NEW YORK • OXFORD
PARIS • SAN DIEGO • SAN FRANCISCO • SINGAPORE • SYDNEY • TOKYO

Elsevier Butterworth-Heinemann
Linacre House, Jordan Hill, Oxford OX2 8DP
30 Corporate Drive, Burlington, MA 01803

First published by Arnold, 1998
Reprinted 2003
Second edition 2005

British Library Cataloguing in Publication Data
A catalogue record for this book is available from the British Library

Library of Congress Cataloguing in Publication Data
A catalogue record for this book is available from the Library of Congress

ISBN 0 7506 5865 7

For information on all Elsevier Butterworth-Heinemann
publications visit our website at http://books.elsevier.com

Typeset by Charon Tec Pvt. Ltd, Chennai, India
www.charontec.com
Printed and bound in Italy

Contents

1   Introduction   1

   1.1   The original version   1
   1.2   What is traffic?   1
   1.3   What is traffic engineering?   2
   1.4   How much traffic?   2
   1.5   The structure of this book   4
   References   6

2   Traffic Surveys   7

   2.1   Introduction   7
   2.2   How to define a traffic survey   10
   2.3   Traffic counts   10
   2.4   Area-wide surveys   14
   2.5   Speed surveys   22
   2.6   Queue length/junction delay surveys   24
   2.7   Video surveys   26
   2.8   ANPR and probe vehicles   26
   References   27

3   Parking Surveys   28

   3.1   Introduction   28
   3.2   Choosing when to survey   28
   3.3   Supply surveys   29
   3.4   Occupancy surveys   29
   3.5   Beat surveys   30
   3.6   Continuous observation survey techniques   35
   3.7   Summary   37
   References   37

4   Estimating Travel Demand   38

   4.1   Introduction   38
   4.2   Growth factor   40
   4.3   Low-cost manual estimation   41
   4.4   Computer-based traffic models   43
   4.5   Micro-simulation   45

| | 4.6 | Accurate and appropriate data | 46 |
| | | References | 48 |

5   Capacity Analyses   **49**

5.1	Capacity definition	49
5.2	Effect of width on capacity	49
5.3	Effect of gradient	50
5.4	Effect of alignment	50
5.5	Design flows	50
5.6	Flow–capacity relationships	52
5.7	Junction capacity	54
5.8	Merges and diverges	58
5.9	Weaving sections	58
	References	58

6   Traffic Signs and Markings   **59**

6.1	Introduction	59
6.2	What is a traffic sign?	59
	References	65

7   Traffic Management and Control   **66**

7.1	Objectives	66
7.2	Demand management	66
7.3	Engineering measures	69
7.4	Junction types	70
7.5	Road markings	73
7.6	Traffic signs	75
	References	75

8   Highway Layout and Intersection Design   **76**

8.1	Highway link design standards	76
8.2	Priority junctions	84
8.3	Roundabouts	86
8.4	Signalled junctions	87
	References	89

9   Signal Control   **90**

9.1	Introduction	90
9.2	Fixed-time control	90
9.3	UTC	92
9.4	Other UTC facilities	101
9.5	Traffic signal capacity assessments	101
	References	103

10   Parking: Design and Control   **105**

10.1	Introduction	105
10.2	On-street parking	105
10.3	Off-street car parking	110
10.4	Whole life care of car parks	111
10.5	Disabled drivers	112

10.6	Servicing bays and lorry parking	112
10.7	Parking control systems	113
	References	121

11 Road Safety Engineering — **123**

11.1	Factors resulting in accidents	123
11.2	Road accident definition	123
11.3	Police recording	124
11.4	The National Accident Reporting System and STATS19	126
11.5	Data transfer to the highway authority	126
11.6	Other data to be collected for safety work	131
11.7	Presenting accident data for analysis	131
11.8	Summarising data	134
11.9	Data presentation and ranking	139
11.10	Statistical analysis	140
11.11	Problem identification	141
11.12	Institution of Highways and Transportation accident investigation procedures	142
11.13	Designing road safety engineering measures	144
11.14	Accident savings	144
11.15	Road safety plans	148
11.16	Road safety audits	148
11.17	Effect of speed	148
11.18	Monitoring performance of remedial measures	149
	References	152

12 Traffic Calming — **153**

12.1	Objectives	153
12.2	Background	153
12.3	Site selection and ranking	154
12.4	Consultation	156
12.5	Traffic calming techniques	156
12.6	Achievable speed reductions	163
12.7	Estimate of accident reductions and benefits	163
12.8	Urban design	163
	References	164

13 Public Transport Priority — **166**

13.1	Design objectives	166
13.2	The reliable track	167
13.3	Bus priority measures	167
13.4	Bus lanes and busways	167
13.5	Traffic and parking management measures	172
13.6	Traffic signal control	173
13.7	Bus stop improvements	175
13.8	Impacts on other users	178
13.9	The process of designing and evaluating bus priority measures	179
	References	179

14 Development Process and Sustainable Development — **180**

14.1	Planning context	180
14.2	Planning reforms	182

14.3	The role of the transport engineer	183
14.4	Traffic Impact Assessment/Transport Assessment	185
14.5	Elements of a TA	190
14.6	Accessibility planning using ACCESSION	193
14.7	Environmental assessment	197
14.8	Sustainable development	199
14.9	Public resistance to sustainable development policies	202
14.10	Responses by transport planners and engineers	203
14.11	Travel plans	204
	References	207

15   Intelligent Transport Systems   209

15.1	Introduction	209
15.2	Using ITS	209
15.3	Traffic management and control	210
15.4	Tolling	212
15.5	Road pricing	212
15.6	Public transport travel information and ticketing	213
15.7	Driver information and guidance	213
15.8	Freight and fleet management	214
15.9	Vehicle safety	214
15.10	System integration	214
15.11	International comparisons	214
	References	217

16   Enforcement   218

16.1	Introduction	218
16.2	Background	218
16.3	Parking, waiting and loading	219
16.4	Bus lanes	220
16.5	Other traffic offences	221
16.6	Congestion charging	221
16.7	Comment	221
	References	222

17   Statutory Requirements   223

17.1	Introduction	223
17.2	The scope of legislation	224
17.3	The Highways Acts	224
17.4	The Road Traffic Acts	225
17.5	The Road Traffic Regulation Act	225
17.6	Other legislation	226
	References	226

Index   227

1
Introduction

1.1 The original version

In 1970, the first version of this book,[1] written by Gordon Wells was published as an introduction to the then relatively new subject of traffic engineering. Since that date the book has been twice updated by the original Author who produced the last volume in 1978. However, in the last two decades the range of skills required by the modern traffic engineer has developed virtually beyond recognition and it became clear to the publishers that the book needed rewriting.

Gordon Wells has now moved on and in 1998 the current authors published a new version of this book. However in a little over 5 years the world of traffic engineering has moved on and an update was over-due. Therefore a new and extended version has been prepared to keep up to date with current practice.

The purpose of this volume has always been to provide the reader with a basic understanding of the range of skills and techniques needed by the modern traffic engineer.

This book is an introduction to the subject and as such cannot be exhaustive. Our objective is to provide an introduction to the skills required and to direct the reader to authoritative source material for a more detailed understanding of the subject. Indeed, techniques are developing so rapidly in some areas that some parts of the book may be quickly be superseded by new techniques, particularly in the appliance of technology, almost before the book is published. However, we, the authors, as practitioners, have sought to set out the basics in a clear and easy to understand style.

1.2 What is traffic?

Traffic can be defined as the movement of pedestrians and goods along a route, and in the 21st century the biggest problem and challenge for the traffic engineer is often the imbalance between the amount of traffic and the capacity of the route, leading to congestion. Traffic congestion is not a new phenomenon. Roman history records that the streets of Rome were so clogged with traffic, that at least one emperor was forced to issue a proclamation threatening the death penalty to those whose chariots and carts blocked the way. More recently pictures of our modern cities taken at the turn of the century show streets clogged with traffic.

What do we mean by traffic in the context of this book? The dictionary describes 'traffic' as the transportation of goods, coming and going of persons or goods by road, rail, air, etc. Often in common usage we forget this wider definition and colloquially equate the word with motorised road traffic, to the exclusion of pedestrians and even cyclists. Traffic engineering is concerned with the wider definition of traffic and this book deals with the design of facilities of most forms of road traffic. Thus we deal with pedestrians, cyclists and motorised traffic including powered two wheelers, cars, buses and trams and commercial vehicles.

We do not deal with animals, although horses and herd animals have the same rights to use a highway as a pedestrian or a motor vehicle. This sort of traffic is excluded for practical reasons. Apart from the countryside the days when herds are driven on the highway are long-gone with animals being moved mostly by motor vehicle for journeys of more than a few hundred metres. Further making provision for horses is a specialised area beyond basic engineering. This book also excludes any reference to railways, other than in the context of on-road light rapid transport and tram systems and also excludes aviation and shipping.

1.3   What is traffic engineering?

In the introduction to his book Gordon Wells quoted the Institution of Civil Engineers[1] for his definition of traffic engineering, that is:

> That part of engineering which deals with traffic planning and design of roads, of frontage development and of parking facilities and with the control of traffic to provide safe, convenient and economic movement of vehicles and pedestrians.

This definition remains valid today but there has clearly been a change in the emphasis in the role of the traffic engineer in the time since this book was first produced. In the 1970s the car was seen as the future and the focus was very much 'predict and provide'. Traffic engineers were tasked with increasing the capacity of the highway system to accommodate what seemed and endless growth in motor traffic, often at the expense of other road users. Road capacity improvements were often achieved at the expense of pedestrian freedom of movement, pushing pedestrians to bridges and underpasses so that the surface could be given over to the car. However, it is now generally, but by no means universally recognised that we will never be able to accommodate unconstrained travel demand by car and so increasingly traffic engineering has become focused on sharing space and ensuring that more sustainable forms of transport such as walking and cycling are adequately catered for.

This change has been in response to changes in both society's expectations and concerns about traffic and the impact of traffic on the wider environment. There has also been a pragmatic change forced on traffic engineers as traffic growth has continued unabated and so the engineer has been forced to fit more traffic onto a finite highways system.

Since 1970, road travel in the UK has increased by about 75% and, although many new roads have been built, these have tended to be inter-urban or bypass roads, rather than new roads in urban areas. Thus, particularly in urban areas, the traffic engineer's role is, increasingly, to improve the efficiency of an existing system rather than to build new higher capacity roads.

1.4   How much traffic?

By the end of the First World War there were about one-third of a million motor vehicles in the UK. Within 6 years this number had increased by a factor of nearly 5 to 1.5 million vehicles. Table 1.1 shows the growth in traffic since then.

The rapid increase in vehicle ownership (+44%) in the last 25 years is clear; Table 1.2 compares growth in the UK with that in other countries. However, the change in the numbers of vehicles does not tell the whole story.

Table 1.1 Growth in UK vehicles[2]

Year	Number of vehicles
1919	330 000
1925	1 510 000
1935	2 611 000
1945	2 606 000
1955	6 624 000
1965	13 259 000
1975	17 884 000
1985	21 159 000
1995	25 369 000
2000	25 665 000

Table 1.2 Domestic versus overseas traffic growth[*]

	Cars and taxis		Goods vehicles[1]		Motor cycles, etc.[2]		Buses and coaches		Total	
	1984	1994	1984	1994	1984	1994	1984	1994	1984	1994
Great Britain	16 775	21 231	1 769	2 438	1 419	757	149	154	20 112	24 580
Northern Ireland	439	509	37	59	16	9	2	5	494	582
United Kingdom	17 214	21 740	1 806	2 497	1 435	766	151	158	20 606	25 161
Belgium	3 300	4 210	232	403	498	187	12	15	4 042	4 815
Denmark	1 440	1 610	239	313	200	49	8	13	1 887	1 985
France	20 800	24 900	2 868	3 606	5 065	2 561	62	79	28 795	31 146
Germany	28 375	39 765	1 497	2 114	4 223	3 750	124	71	34 219	45 700
Greece	1 151	1 959	572	808	154	388	18	23	1 895	3 178
Irish Republic	717	939	84	136	26	24	3	5	830	1 104
Italy	20 888	30 420	1 683	2 543	5 163	5 397	72	78	27 806	38 438
Luxembourg	146	229	9	13	3	8	1	1	159	271
Netherlands	4 841	5 884	345	565	785	844	12	12	5 983	7 305
Portugal	1 265	3 532	347	442	102	192	10	11	1 724	4 178
Spain	8 874	13 441	1 643	2 832	706	1 279	41	47	11 264	17 599
Austria	2 468	3 479	203	283	646	742	9	10	3 326	4 514
Croatia	–	698	–	42	–	9	–	9	–	759
Czech Republic	2 640	2 967	260	161	612	476	35	23	3 547	3 627
Finland	1 474	1 873	171	249	211	158	9	8	1 865	2 288
Hungary	1 344	2 179	142	245	394	157	25	22	1 905	2 602
Norway	1 430	1 633	90	319	176	161	16	29	1 712	2 142
Slovak Republic	–	994	–	148	–	229	–	12	–	1 383
Sweden	3 081	3 594	210	304	22	57	14	14	3 327	3 969
Switzerland	2 552	3 165	193	250	847	708	12	14	3 604	4 137
Japan	27 144	40 772	17 010	22 246	17 354	16 396	230	248	61 738	79 662
USA	1 27 867	1 33 930	38 047	63 445	5 480	3 718	584	670	1 71 978	2 01 763

* Numbers represent thousands.

Table 1.3 Change in road traffic in the UK[2]

Year	Estimated vehicle (km $\times 10^9$)
1975	244.4
1985	309.7
1995	430.9
2000*	467.7

*The year 2000 figures were affected by the September fuel protests, the figure for 2001 is 473.7.

Table 1.4 Change in the highway network in the UK[2]

Year	Network length (km)
1985	348 699
1995	366 999

Not only do more people own cars, each vehicle is used more. In addition, the pattern of freight movement has changed dramatically with a shift from rail to road and radical changes in distribution procedures, which mean that goods now tend to be distributed from fewer, larger depots, with a consequent increase in goods vehicle travel. Table 1.3 shows the increase in travel on the roads system since 1975.

During the same period, the highway network was increased by the construction of new roads, such as, bypasses and, latterly by motorways. Table 1.4 shows the increase in network size, between 1985 and 1995.

In the UK, the Department of Transport uses a method of predicting future traffic which links growth in car ownership to, among other things, predicted changes in economic performance.[3,4] The methodology provides high and low predictions based on expectations about economic performance. The predictions, which have, if anything, proved to be historically conservative, suggest that, within the next 25 years traffic levels will more than double.

1.5   The structure of this book

This book has been prepared in the context of traffic engineering as practised in Britain, with references to standard UK designs and legislation. That said, many of the basic principles are the same regardless of the country in which they are applied. The authors hope therefore that the information contained in the book will be of general interest to a wider audience who will be able to use and adapt the information contained in this book to their circumstances.

The book is set out to follow a logical sequence of steps designed to allow the reader to first measure and understand 'traffic' and then to design measures to deal with and control it.

Chapters 2 and 3 deal with surveys. Before attempting to undertake any task, it is important to obtain a measure of, at least, the base level of traffic, to ensure that any solution is appropriate and of correct scale. Chapter 2 describes a range of survey methods for measuring traffic flow on a road system while Chapter 3 deals with parking surveys.

Sometimes a solution can only be implemented in response to existing conditions. For example, a car park control scheme may be appropriate for today's conditions but may become obsolete as circumstances change. It is not always possible to predict the future and so, for some problems, it is only appropriate to design a scheme, which deals with the existing problem.

Therefore, an understanding of existing traffic conditions is fundamental. Often however it is equally important to ensure that any traffic engineering solution will be capable of dealing with both existing traffic and the traffic expected throughout the design life of the project. For example a junction design should take account of both present-day flow and expected growth for an agreed period. Indeed, the basis of traffic prediction of future traffic, from a new development is a fundamental part of the transport impact analysis process. The techniques for predicting traffic flow are dealt with in Chapter 4.

Chapter 5 deals with the concepts of traffic capacity and provides definitions of capacity and flow. The chapter describes techniques for estimating the effects of traffic flow on junction and highway performance.

This is followed by a number of chapters which deal with the fundamentals of traffic management and control, covering issues such as signing and the design of roads and junctions, and traffic signals. Increasingly in trying to allocate priority to different types of traveller and vehicle we allocate specific parts of the road to a particular type or group of vehicles, this is covered in Chapter 9. Chapter 10 deals with vehicle parking, both on- and off-street.

Road safety is a very important issue and a traffic engineer has a duty to ensure that any work he undertakes will result in a safe environment for all road users. Chapter 11 deals with road safety engineering and discusses concepts of designing in safety and measures to deal with pre-existing problems. This includes a description of the safety audit procedure, which should now be a mandatory part of any new scheme.

Chapter 12 deals with traffic calming, describing how traffic engineering measures can be used to manage and reduce the adverse impacts of vehicular traffic to provide a safer and better environment in existing streets. This chapter also deals with priority measures for public transport, including measures for the integration of light rail systems into road traffic.

Road-based public transport, that is buses and light rail (trams) plays a very important role in movement in our towns and cities and Chapter 13 introduces the various techniques available to meet this need and give priority to these more efficient forms of transport.

Any new land-use development, be it a housing estate, a new factory or an edge of town shopping centre, will generate traffic as people travel to and from the site. Chapter 14 discusses the current government view on development and sustainable transport and the techniques used to estimate the amount of traffic at different types of development, both during construction and when the development is complete. The chapter also describes how to measure both the community and environmental impacts that arise.

This chapter also develops the concept of sustainable transport further, dealing with the development control process and discusses measures designed to help promote cycling and walking and to minimise total travel demand.

The word telematics has gained common usage in recent years, often used as a talisman to describe anything technical which has an application to traffic and travel. Unfortunately many of the ideas, which are so enthusiastically promoted by their developers suffer from inadequate development or are frankly solutions in search of a problem. Chapter 15 gives a sound basic grounding of how new technology is being applied successfully to improve the efficiency and safety of the transportation system, including a review of ideas which are likely to gain widespread use in the next few years.

Chapter 16 deals with the rapidly growing area of enforcement. Traditionally the sole preserve of the Police, as traffic control becomes more extensive and more complex we are rapidly moving towards an environment where the responsibility for enforcing minor regulations is moving away from the Police to the highway authority, with greater reliance on the use of technology to detect and penalise transgressors. Finally Chapter 17 gives a brief overview of the current UK legislative framework under which the traffic engineer operates.

References

1. Wells, GR (1976) *Traffic Engineering An Introduction*, Charles Griffin & Company.
2. Department of Transport (1996) *Transport Statistics Great Britain*, HMSO, London.
3. Department of Transport (1996) *National Road Traffic Forecasts, 1989, Rebased at 1994*, HMSO, London.
4. Department of Transport (1989) *Highways Economic Note 2*, HMSO, London.

2
Traffic Surveys

2.1 Introduction

Traffic engineering is used to either improve an existing situation or, in the case of a new facility, to ensure that the facility is correctly and safely designed and adequate for the demands that will be placed on it.

In an existing situation we have to know the present-day demands and patterns of movement, so that the new measure can be designed adequately. With a new road or facility, there is obviously no existing demand to base the design on; therefore, we have to estimate the expected demand.

If a new facility replaces or relieves existing roads, for example a bypass or a new cycle track, we can estimate the proportion of traffic that could be expected to transfer using a traffic assignment (see Chapter 4).

If the facility is completely new, for example a road in a new development, then the expected traffic and hence the scale of construction needed has to be estimated another way. This is usually done by a transport assessment (see Chapter 14) which will seek to assess the likely level of traffic by reference to the traffic generated by similar developments elsewhere. In either case the starting point will be a traffic survey.

The main reason for undertaking a traffic survey is to provide an objective measure of an existing situation. A survey will provide a measure of conditions **at the time that the survey was undertaken**. A survey does not give a definitive description of a situation for ever and a day and if the results are to be used as representative of 'normal' traffic conditions, the survey must be defined with care and the information used with caution.

Traffic flow varies by time of day, day of the week and month of the year. Figure 2.1 shows a typical 24-hour daily flow profile for an urban area. The figure shows morning and evening peaks as people travel to and from work. Flow drops off at night, to a lower level than observed either during the day, when commercial activity takes place, or in the evening, when social activities tend to take place.

Traffic flows also tend to vary by day of the week (Figure 2.2). Again, on a typical urban road traffic flows tend to build during the week, to a peak on Friday. Flows are lower at the weekend, when fewer people work and lowest on Sunday, though the introduction of Sunday trading has affected the balance of travelling at the weekend.

The variation in pattern of travel over the year depends a great deal upon location. In urban areas, which are employment centres, flow drops during the summer period when schools are closed and workers tend to take annual holidays. This is balanced by a reverse trend in holiday areas, where traffic flows increase dramatically in July and August, and roads which are adequate

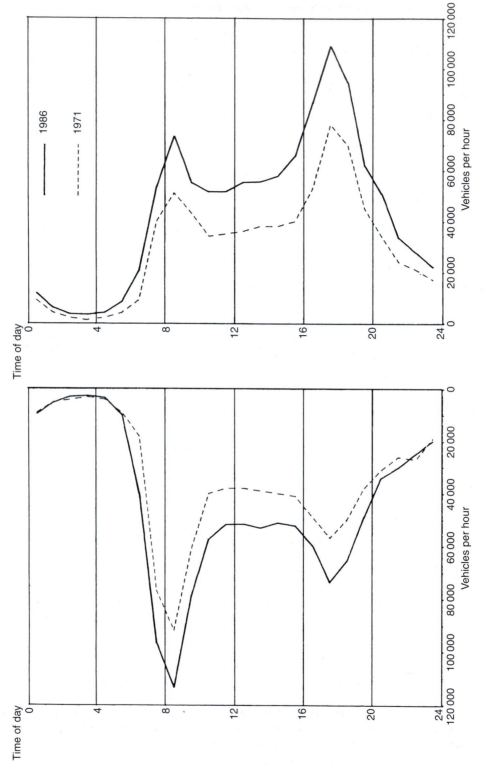

Fig. 2.1 Graph showing 24-hour flow profile.

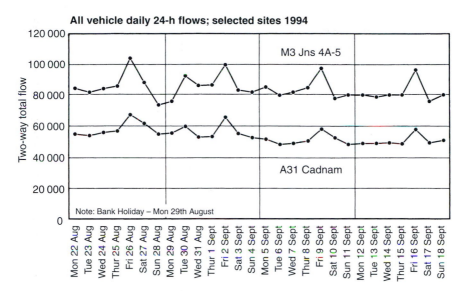

Fig. 2.2 Variation in flow by day of the week: all vehicle daily 24-hour flows for selected sites in 1994. Note that 29 August was a bank holiday.

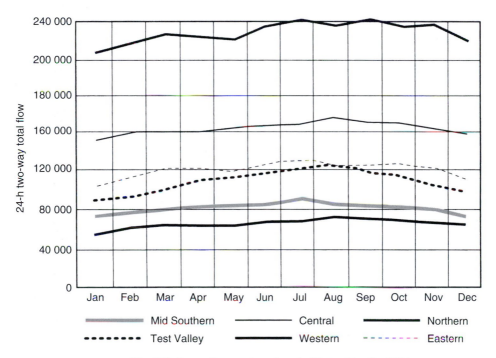

Fig. 2.3 Screenlines and cordons in Hampshire for 1994.

most of the year become heavily congested. The effect can be less dramatic on inter-urban roads, other than those providing access to holiday areas, as, to an extent, the decline in inter-urban business travel during the summer is off-set by tourism. Figure 2.3 shows examples of typical annual flow profiles for roads in areas of Hampshire.

The information above shows that the pattern of flow on any road can be highly variable and, in deciding when and where to undertake a traffic survey, it is important to take care to ensure that the survey provides a fair measure of the traffic conditions that are being studied. To take the example of the road in a tourist area, a traffic survey on an August bank holiday would measure peak traffic conditions. As these levels occur only 1 or 2 days a year there would be little point in using this data as a basis for design, as the scheme would be over designed for traffic conditions most of the time.

Generally, traffic surveys should not be planned to measure the 'peak of the peak' but to measure the 'normal' peak conditions. Trunk road surveys may require a full year's survey of traffic so that the 50th (30th or 200th) highest hourly flow can be determined, and used as the basis for design.

2.2   How to define a traffic survey

The starting point in defining a traffic survey is to decide what question has to be answered and choose the type of survey accordingly. If the survey is not adequately planned, there is a danger that the wrong data will be collected and the traffic situation will not be correctly understood.

The only exception to this rule occurs when one is faced with a complex situation where it may not be possible, at first, to adequately understand what is going on, in terms of traffic flow and circulation. In these circumstances the traffic survey is providing evidence which will not only be used to quantify behaviour, it may also be used to define it.

2.3   Traffic counts

The traffic engineer has an increasing number of survey methodologies available to help him to understand traffic movement. The main techniques are described below, with their principal applications. All the traffic count methodologies described are non-interventionalist, that is they do not affect the traffic flow being measured.

2.3.1   AUTOMATIC TRAFFIC COUNTS

Automatic traffic counters are used to mechanically measure traffic volumes moving past the survey point. The counters normally use a pressure tube or an inductive loop which is fixed across the road at the census point.

A pressure tube is compressed each time a vehicle axle crosses it. This sends a pulse along the tube which is counted and hence the vehicular flow can be estimated. More modern systems use a piezzo electronic tube and the electrical pulses are counted. Figure 2.4 shows a typical output from an automatic traffic counter with the data presented as hourly flow.

A tube counter measures the impact of an axle and so traffic flow is derived from counting the number of impulses and dividing them by a factor representing the most common number of axles on a vehicle (i.e. two). On heavily trafficked roads where there are large numbers of multi-axle heavy vehicles, a slightly higher factor may be used. Inaccuracies can occur when two vehicles cross the loop at the same time, for example a motorcycle and a car, or when there is a higher than expected proportion of multi-axle vehicles. In high speed conditions, 'axle bounce' can also mean that an axle bounces because of the road surface conditions and fails to compress the tube.

An alternative is to use an inductive loop which will detect the mass of a vehicle. The passage of the metal mass of a vehicle over the loop induces a magnetic field in the loop, allowing the presence of a vehicle to be registered. This type of technology counts the vehicles' presence

Week beginning: Thursday 24 April 1997
Site reference: CROY1
Vehicle flow: CHANNEL 2 – Outbound

Hour ends	Thursday	Friday	Saturday	Days Sunday	Monday	Tuesday	Wednesday	5 day average value %		7 day average value %	
1	4	4	18	17	5	0	1	3	0.83	7	1.89
2	0	1	3	4	0	3	0	1	0.24	2	0.43
3	3	0	0	3	0	0	0	1	0.18	1	0.23
4	0	0	1	1	1	0	1	0	0.12	1	0.15
5	0	0	0	0	1	2	0	1	0.18	0	0.12
6	2	0	2	0	0	0	0	0	0.12	1	0.15
7	2	3	1	0	4	5	5	4	1.12	3	0.77
8	8	6	2	3	24	18	14	14	4.13	11	2.90
9	37	20	5	25	11	32	24	25	7.32	22	5.95
10	11	7	8	39	11	13	6	10	2.83	14	3.67
11	7	6	38	49	7	9	9	8	2.24	18	4.83
12	13	12	10	33	5	10	8	10	2.83	13	3.52
13	8	14	25	34	9	16	8	11	3.24	16	4.41
14	11	11	15	44	10	15	14	12	3.60	17	4.64
15	11	10	10	44	6	11	5	9	2.54	14	3.75
16	32	16	9	46	12	24	11	19	5.60	21	5.80
17	23	11	18	36	10	11	7	12	3.66	17	4.48
18	34	12	21	47	11	33	17	21	6.31	25	6.76
19	76	15	39	26	39	51	35	43	12.74	40	10.86
20	69	17	29	25	36	59	35	43	12.74	39	10.44
21	50	22	81	15	24	65	12	35	10.21	38	10.40
22	14	59	18	4	19	37	25	31	9.09	25	6.80
23	44	10	16	4	3	15	24	19	5.66	17	4.48
24	6	21	14	10	1	6	8	8	2.48	9	2.55

Totals

	Thursday	Friday	Saturday	Sunday	Monday	Tuesday	Wednesday	5 day	7 day
7–19	271	140	200	426	155	243	158	193	228
6–22	406	241	329	470	238	409	235	306	333
6–24	456	272	359	484	242	430	267	333	359
1–24	465	277	383	509	249	435	269	339	370

Fig. 2.4 Example of automatic traffic counter output.

directly with one pulse for each vehicle. The loops can give false readings if two vehicles pass the loop simultaneously or are close together; conversely a vehicle pulling a trailer can be read as two vehicles.

Automatic counters can also be set up to classify the type of vehicle by numbers of axles.

Automatic traffic counters are usually used where traffic flow data is required over an extended period, for example a week or a year. The data can be presented in terms of the flow per time period, for example per hour, per day or per week and used to compare daily, weekly or seasonal variation, as well as quantifying the volume of traffic. Automatic counters are useful when one wishes to collect indicative data over an extended period cheaply. The counters cannot give precise information.

For a counter which has been installed at a particular location for a short period of time, the data can be collected at the side of the road and stored on a data tape which can be collected periodically

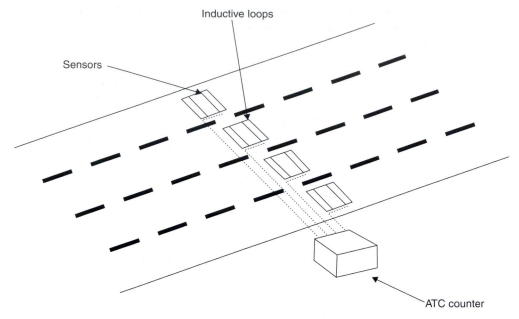

Inductive loops

Sensors

ATC counter

Fig. 2.5 Typical automatic traffic counter installation.

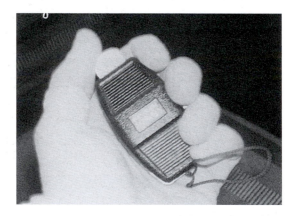

Fig. 2.6 Tally counter.

(Figure 2.5). If the counter is part of a permanent installation the data can be collected remotely using telemetry.

2.3.2   *MANUAL COUNTS*

Traffic flows can be measured by manual observation, instead of using an automatic counter. Traffic flowing past a survey point is counted by an observer, who would record the flow using either a tally counter (Figure 2.6) or by taking a manual count of vehicles and recording it on paper, typically using a five bar gate counting technique, or by using a hand-held computer.

Counts are classified, to identify the volume and mix of types of vehicles using the road at the survey point. Figure 2.7, from the DOT publication *Roads and Traffic in Urban Areas*,[1] shows a

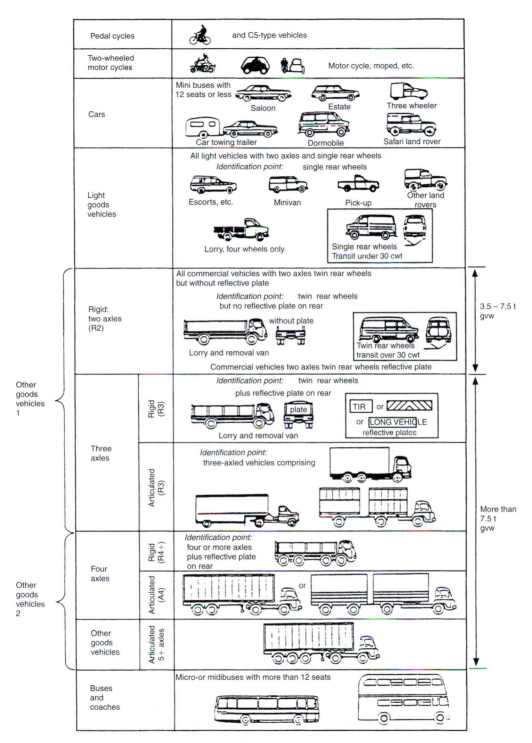

Fig. 2.7 Vehicle categories used for survey purposes.

typical classification. However, the level of classification used will very much depend upon the needs of the survey. For example, it may be adequate to use a simpler form of classification, such as cars and taxis, buses and commercial vehicles. The engineer should choose an appropriate level of classification for each study.

If a data collection survey is only planned to cover a short period of time, then the expense of installing an automatic counter may not be justified when compared with the cost of using a surveyor. The surveyor also has the ability to discriminate between classes of vehicles.

Manual counts generally offer better value for money when data is to required for a single day or for less than the full 24-hour day but collected over 2 or 3 days. Manual classified counts (MCCs) become more difficult where flows are very high, and where any break in concentration can introduce high error rates in the count. Figure 2.8 shows a typical survey form for an MCC.

If the engineer wishes to gain a quick insight to traffic conditions over a wider area, short period, sample traffic counts can be taken over a wide area and factored up, to represent the hourly flow. Thus, for example, if one wished to have an understanding of traffic levels at a complex junction, traffic could be counted at each arm for 5–10 minutes and then factored up to hourly counts, to give an understanding of conditions. This is a good method of gaining a quick insight into traffic levels but should not be used as a substitute for a properly organised traffic survey.

2.3.3   TURNING MOVEMENTS

A manually-classified count (MCC) records directional traffic flow past a survey point. The survey point could be mid-link or at a junction. If we wish to understand how traffic is behaving at a junction more precisely, we extend the complexity of the MCC to include a measure of turning movements. Thus at a four arm junction, surveyors would record both the flow and the direction of turn (Figure 2.9). This sort of data would typically be used to analyse the traffic conflicts at a junction, and to determine whether or not the junction needed to be modified.

Once again the count can be classified to identify the mix of traffic. This can be very useful as different types of vehicle have different acceleration, turning and braking characteristics, which will affect the amount of traffic that can pass through a junction.

2.4   Area-wide surveys

The surveys described above are adequate for measuring traffic flow and direction of movement at a single point, or at a single junction. However, if we wish to understand movement over a wider area, then other methods have to be used. Three techniques are described below; one for numberplate surveys and two covering origin and destination (O&D) surveys.

2.4.1   NUMBERPLATE SURVEYS

We may wish to understand how traffic is circulating in a limited area. This could be for example, a complex gyratory system, a residential area where we suspect that there may be 'rat-runs', or even a town centre ring road where we wish to understand if traffic uses the ring road or passes through the town centre.

The technique used is to record the registration mark of each vehicle as it enters and leaves the system being studied and then to match the registration marks, to establish how a vehicle travelled through the road system being studied. It is not normally necessary to record the full registration mark.

If the first four characters of a typical UK registration plate are considered, then the probability of having the same four first characters on a different vehicle is the product of the probability

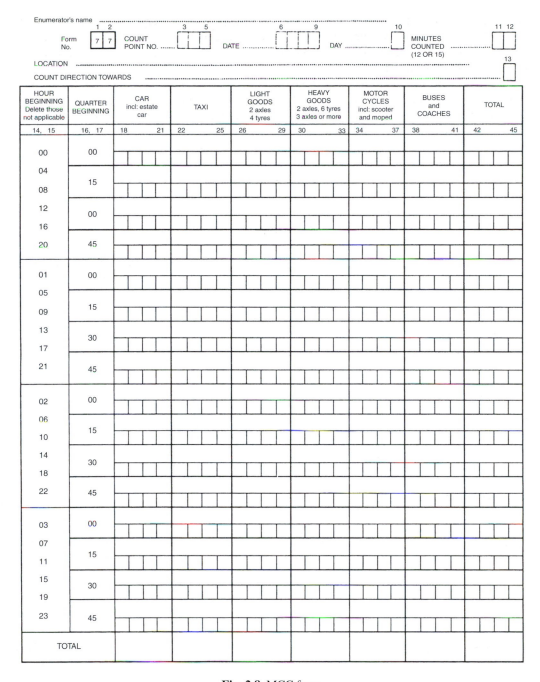

Fig. 2.8 MCC form.

of these same four characters appearing. This risk is further reduced by the chance of the vehicles being the same type.

It can be seen from the above that, in practice, it is sufficient to record the first four characters of each registration mark as, in all but the largest systems, the chances of getting two or more vehicles with the same four first characters are insignificant.

Location diagram:		Enumerator		Job no:				Time start:		
		Date:		Location:				Time finish:		
		Weather:		Day:				Comments:		
Turning movement / Vehicle type										
Pedal cyclist										
Motor cyclist										
Cars/ LGVs										
HGV 1 (2 – 3 Axles)										
HGV 2 (4 Axles)										
All bus types (including mini buses)										

Fig. 2.9 Turning count form.

In theory this is a very simple and robust survey technique. In reality, it suffers from a number of practical problems. The first of these is survey error. Even on the best run survey it is likely that 5–10% of registration numbers will be unmatched, due to errors in reading or recording numberplates.

For a complex situation, for example where there is more than one route between an entry and exit point, the survey could require data to be recorded at an intermediate point, as well as the entry and exit points. This adds to both the complexity of data collection and matching and analysis.

Obviously a vehicle cannot leave before it arrives and to help prevent spurious matches the time that a vehicle is observed should also be recorded. This data also provides approximate journey-time information.

When the survey covers an extended area, for example a rat-running survey, then vehicles, such as those belonging to local residents may enter the survey area and stop, or start within the survey area. A vehicle may also enter the study area, stop for a while and then leave.

To ensure that one understands what the survey results represent, the survey has to be carefully specified to take account of these factors. Thus, for example, if a large number of vehicles are expected to leave and/or join the traffic flow within the survey cordon, high levels of mismatch can be expected. If however the system is closed, for example a gyratory system, then there should be a very high match.

2.4.2 ORIGIN AND DESTINATION (O&D) SURVEYS

The alternative way to establish where drivers are travelling is to ask them, using an O&D survey. Various types of O&D surveys are used as a part of the wider transport planning process. However, this is beyond the scope of this book and is not explored here. The standard techniques are roadside interview surveys and self-completion questionnaires.

In most cases it will be impossible to carry out a 100% survey of drivers and so we must rely on a response from a sample of drivers in the traffic flow. Clearly, if the survey results are to be relied on, the sample should be unbiased with all types of vehicles and movements represented.

2.4.3 ROADSIDE INTERVIEW SURVEYS

At a roadside interview survey, a sample of drivers is stopped at the side of the road and asked their O&D, plus any other data which could of relevance, such as journey purpose. Figure 2.10 shows a typical survey form.

The size of sample will depend on flow and the level of reliability required. This is described in greater detail in Traffic Advisory Leaflet TA 11/81.[2] However, the theoretical advice offered in this guidance has to be balanced by what can be practically achieved. If an interview lasts for a just a minute and after allowing time for the driver to enter and leave the interview bay, the time stopped is say 2 minutes, then each interviewer could handle 30 drivers an hour. Simple logic dictates that there has to be a limit on the length of an interview bay, for practical reasons, if not as a result of the road's geometry, and this will determine the absolute number of drivers that can be interviewed each hour.

Typical designs for interview stations are shown in Figure 2.11. Where surveyors are having to work close to moving traffic, the safety of all involved is paramount.

Traffic flow is directed past the interview point and a sample of vehicles is directed into the interview bays where the drivers can be asked about their journey. The power to direct traffic resides only with the police and so these types of surveys require the cooperation and continuous presence of a police officer.

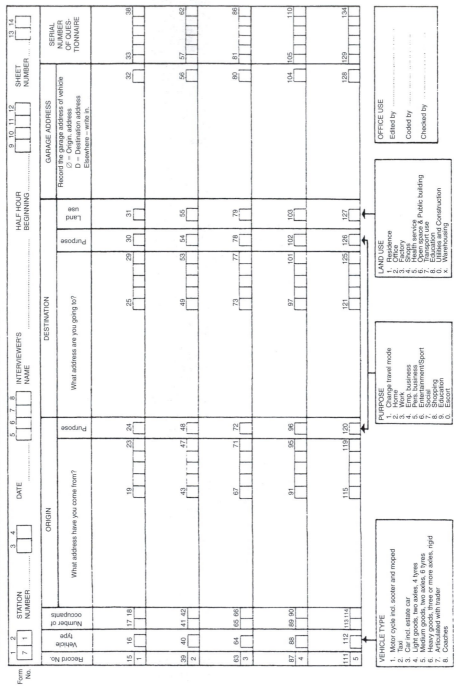

Fig. 2.10 Roadside interview form.

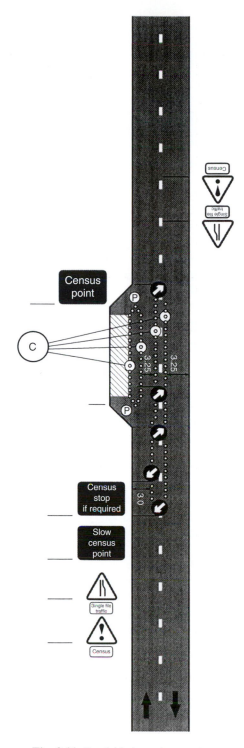

Fig. 2.11 Roadside interview station.

As the direction of traffic at a survey station requires the presence of a police officer, it is important to involve the police in the design of the survey, to ensure that they are satisfied that the survey can be conducted safely and that the officer(s) involved are aware of the need to gain a representative sample from the traffic flow.

Once a driver has been selected for interview and is stopped in the interview bay, he should be asked to provide the necessary answers and then released as soon as possible. Although a driver must stop when instructed to do so by a police officer, there is no obligation on the driver to participate with the interview and a driver may refuse to answer any questions.

The data are used to reconstruct the pattern of vehicular movement by aggregating trip O&D into a pattern of zones and then grouping together the trips to construct a matrix of movements called an origin–destination matrix. The exact grouping of information will depend on the road network and the distribution of developments served by the network. Obviously the boundaries of zones must be chosen so that as far as possible, trip movements can be distributed correctly on the network.

As the interviews represent a sample of traffic, the survey responses have to be factored up to represent the total flow at the survey point. Normally this is done by undertaking a contemporaneous classified traffic count at the survey point and factoring up the sample results to the total flows. This process is known as sample expansion.

Sample expansion is normally achieved by factoring the sample in a given time period, say an hour or 30 minutes, to the observed flow, normally subdivided by vehicle type. Alternative time periods, such as morning and/or evening peak may be used.

It may be difficult to carry out a survey without so disrupting traffic that a contemporaneous count will be unrepresentative of normal traffic conditions. In these circumstances, the pragmatic approach would be to carry out a traffic count on the same day in the preceding or following week and to factor the survey results to these counts.

2.4.4   SELF-COMPLETION FORMS

In some locations, often in congested urban areas, it is not practical to set up an interview station. This could be because road geometry means that it is not possible to safely slow down and stop traffic, or because the volume of traffic means that an unacceptable level of traffic congestion would arise if road space were allocated for a survey bay. In these circumstances the reply paid questionnaire may offer a suitable alternative methodology, to provide the information required.

In most places it should be possible to find a location where traffic flow can be stopped for a short period. This could be a 'natural' interruption, such as at traffic signals, or as a result of a police officer stopping the flow periodically for a short time.

If traffic can be stopped, surveyors can enter the traffic stream and hand out a reply paid card which asks the driver the same questions as would have been asked at a roadside interview. The card can be completed by the driver at their own convenience and posted back to the organisers.

As the surveyors do not have to ask questions, many more cards can be distributed using the same resources as would be used for an equivalent interview survey.

Where there are multiple lanes of traffic it is important to ensure that the distribution of questionnaires between lanes is balanced. The rationale for this is self-evident. If there were three lanes of traffic at a survey point, with one turning left, one going ahead and one going right, then any bias in distribution could produce a bias in response, leading to a misrepresentation and a misunderstanding of the existing traffic flow.

The technique allows contact with a larger proportion of drivers in the traffic stream than would be possible with a roadside interview. The key disadvantages of the method are:

- the response rate cannot be judged in advance and can be highly variable;
- the sample size can vary greatly, with a very low response in some time periods and higher returns in others;
- the sample may not be representative of all time periods and vehicle types;
- the respondents are self-selecting and this may introduce a bias.

With the simple example of a three lane road, if no survey forms were handed out in the right-turning lane then these movements would not be represented and, once the data had been processed and analysed it would prove impossible to use the survey results to reproduce the observed situation. If the issue of questionnaires is monitored and controlled however, this situation can be monitored and the bias avoided.

If the questionnaires are numbered, then the sample response in each time period can be judged. If flow in one period is under-represented in the response, or a particular category of flow is not fully represented, then it is possible to correct for this lack of data by a process of data patching. In simple terms this means that where there is an inadequate sample in one time period, the data from adjoining time periods are combined to allow representation of the traffic movement.

Data patching should only be attempted within carefully defined limits when traffic characteristics can be expected to be similar. Thus for example, it would be acceptable to match successive peak periods when traffic flow is dominated by the journey to work. However, it would be wrong to merge peak and off-peak traffic.

The response rate to such surveys can vary tremendously and unpredictably. We have experienced responses as low as below 20% and above 50% for surveys which are similar in terms of the type of questions asked and the purpose of the survey.

It is now commonplace to offer an incentive to drivers to encourage a higher response rate. This would typically be a prize draw for a cash sum, or a free holiday or gift. There is no conclusive evidence that incentives regularly result in a higher response rate. However, there is no evidence that they deter responses and, on balance, they would seem to be more likely to help than hinder.

As with an interview survey, it is essential to carry out a traffic count so that the results can be factored up. Although the survey technique is designed to have the minimum impact on traffic flow, it is likely that there will be some impact and so it is probably best to plan to record unobstructed traffic flows on another day.

The potential for this type of survey to disrupt traffic was graphically illustrated by a survey at traffic signals in West London designed to capture traffic using the M4. The peak hour survey had surveyors handing out questionnaires to drivers at a signal stop line while the lights were red. The signal settings were unchanged and the extra delay to drivers was caused by a police officer who held the traffic on red/amber to ensure that surveyors were clear of the traffic. The survey was abandoned after about an hour, by which time there was a 20-kilometre tailback on the motorway. In planning any survey, it is important to ensure that the planning takes account of the likely effects on traffic and seeks to minimise any adverse effects.

Occasionally it is not possible to survey traffic at the point where the information is required. With the benefit of hindsight, the A4 survey mentioned above, was one such place. In these circumstances, a more time-consuming and expensive approach has to be adopted. For example, if an engineer wishes to understand the flow on a motorway link, it is not possible to set up a survey on the motorway. The technique adopted in these circumstances is to set up a series of interview stations on motorway accesses upstream of the part of the motorway which is of interest.

Self-completion forms can also be used to provide O&D information on bus passengers; the forms are distributed on the bus by surveyors. Once again, it is important to record both total passenger numbers and the time when each form was handed out, so that replies can be factored up to represent the full travelling population.

2.5 Speed surveys

There are two basic techniques for measuring the speed of traffic. The first method uses speed measuring equipment, such as a radar gun, to record the speed of traffic, or a sample of traffic passing a particular point in space based on the Doppler effect of the change in frequency of the microwave beam reflected by the vehicle. An alternative is the 'time of flight' system using two loop or piezo sensors situated close to one another and measuring the time the vehicle takes to pass from one sensor to the next. The second technique relies on a vehicle travelling in the traffic flow, where the speed is calculated as the time taken to travel a certain distance.

The first of these measurements is called the **spot speed** for an individual vehicle. Spot speed measurements can be used in combination to show the variation of vehicular speeds, as a simple frequency graph (Figure 2.12). Alternately spot speed measurement can be used to calculate the **time mean speed** of traffic passing the measuring point.

Time Mean Speed is the average speed of vehicles passing a point over a specified time period and is defined as:

$$V = \sum_t \frac{V_t}{n}$$

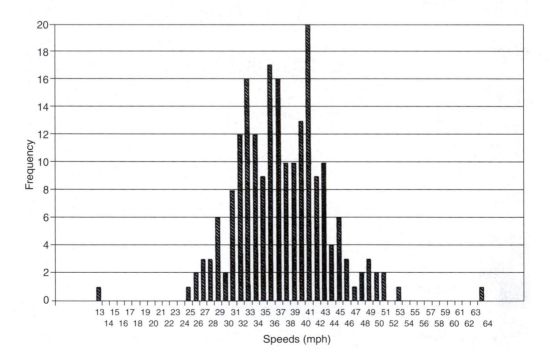

Fig. 2.12 Time mean speed.

where V is the time mean speed,
           V_t is the speed of an individual vehicle,
           n is the number of vehicles observed.

The data required can be collected using a speed measurement device, such as a radar gun, suitably positioned to take readings of the traffic stream it is desired to study. This is illustrated in Figure 2.13.

The second measure of speed is **space mean speed**. This is a measure of the speed of travel over a measured distance, rather than at a single location. Thus if an observer wishes to know the speed of vehicles travelling along a length of road length l, then if each vehicle i takes time t_i to travel the link then space mean speed is defined as:

$$V = \frac{l}{\Sigma_i \dfrac{t_i}{n}}$$

where V is the space mean speed,
           t_i is the travel time of the ith vehicle,
           n is the number of vehicles observed.

Fig. 2.13 Hand-held digital radar gun.

Space mean speed (or journey speed) within a network is commonly measured using what is known as the floating car method. The technique measures average journey time between two locations in a network, along a pre-determined route.

With the floating car method, a car is driven at the average speed of the traffic. This is achieved by driving the car so that it overtakes as many cars as overtake it. Hence the name floating car, as the vehicle 'floats' in the traffic stream, moving at the average speed of the traffic on the network.

Clearly, over an extended journey this may not be too difficult to achieve, however on a short link, or in a busy street, overtaking opportunities may be limited and so the technique may be impractical.

The technique allows an extended survey of speeds throughout a network, with limited survey resources and can be very useful in gaining a broad understanding of traffic speeds in an area. It should be noted that, where average traffic speeds exceed the speed limit this technique is, strictly speaking, inappropriate as the observers would have to speed to collect the data.

Journey speed surveys are often used to collect time series data on highway networks to show how traffic speeds are changing with time. Data is usually collected by recording the journey time in stages, between successive major junctions say, with pre-determined timing points on the vehicle's journey. Obviously time on a single link can vary for a number of reasons. These could include:

- random variations in traffic flow;
- time of day;
- the turning movement the vehicle makes at a junction;
- incidents, such as road works or an accident;
- for traffic signals, where the vehicle arrives in the signal's cycle.

To obtain an average value, it is essential to repeat the journeys a number of times to give data from a range of traffic conditions. Obviously any survey data collected where it can be established that the traffic flow was affected by an accident or road works should be discarded, unless of course the purpose of the survey was to measure the disruption effect of the incident. Pragmatically, the number of runs is likely to be influenced by the budget available for the survey, but a minimum of five good runs should be undertaken, although as few as three have been used in large scale surveys such as the Greater London Speed Surveys.[3]

An alternative to the floating car method is to employ time-synchronised video cameras at locations along a route and record journey times for each vehicle using Automatic Numberplate Reading (ANPR) techniques.

2.6   Queue length/junction delay surveys

Queue length surveys involve recording the length of the queue on an approach to a junction. Typically, the position of the back of the queue is recorded for each lane every 5 minutes at a roundabout and at signalised junctions approximately every 5 minutes as the maximum queue during the current cycle. A typical form of presentation of the resulting information is shown in Figure 2.14.

Junction delay surveys involve one surveyor recording the times and registration numbers of vehicles joining the end of the queue and a second surveyor recording the times and registration numbers of vehicles passing through the junction. By matching registration numbers, very accurate measures of average delay and the standard deviation of delay can be obtained.

Queue length survey

Site : **Stratford Road / Highgate Road**
Direction : **Southbound**

Day: **Wednesday**
Date: **22nd May 1996**

Key Lane1 Inside/ Hubside Lane ——————
Lane2 Offside Lane - - - - - - - -

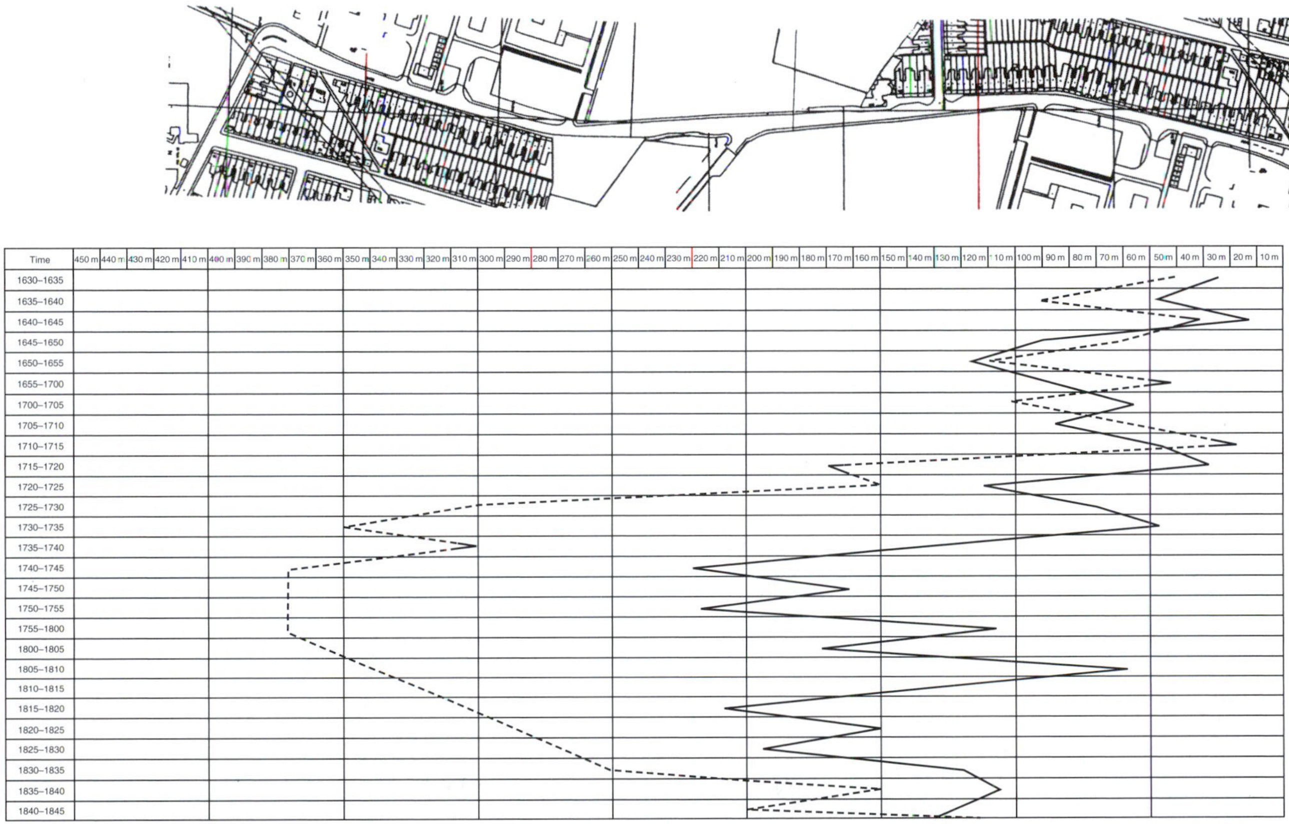

Fig. 2.14 Queue length/junction delay survey.

2.7   Video surveys

The use of video as a data collection tool in traffic engineering is a relatively new but potentially very powerful concept. A strategically placed camera can be used to observe traffic and parking activity in a street and, depending on location and equipment, it is possible to survey up to 400 metres of road from a single vantage point.

Cameras are mounted high to minimise the obstruction of the longer view from vehicles near to the camera. One of the key advantages of a camera is that it records everything that happens. Other survey techniques inevitably only record partial data, collecting just those aspects of traffic behaviour which the survey is designed to record. With a video survey, it is possible to review the video and observe other activities which were thought to be unimportant when the survey was planned.

The video has a particular advantage when flows are very high and it is difficult to count manually, or when we wish to study a particular location where we are not absolutely sure what the key issue is. With a video we can simultaneously record:

- traffic flow
- turning movements
- speeds
- congestion and delays
- parking and loading
- pedestrian movements.

Most importantly, a video allows us to see the interaction of all these factors.

The video also offers the unique advantage of allowing us the opportunity to view the situation repeatedly, until we are satisfied we understand what is happening.

Video surveys are not cheap. Although the data collection may only require the presence of a single technician, to monitor the equipment, the subsequent recording and analysis of data from the video can take up to six times as long as the real-time recording; depending on what data is to be collated from the video and whether or not computer-assisted techniques are available. The authors use a system where the video is displayed under computer control so that an operator can respond to data-entry requirements and readily key information directly into a database. Video survey techniques employing numberplate matching with split-screen presentation of the data for the analyst can also be used as a substitute for manual methods of undertaking journey time and area-wide surveys.

2.8   ANPR and probe vehicles

ANPR systems are now being commonly used for measuring traffic speeds by recording at two camera sites the times of vehicles passing and their registration numberplate. From this information the speed profile of vehicles over a section of road can be determined.

Instrumented Probe vehicles can provide spot speeds and are equipped with:

- on-board equipment for continuous measurement of the vehicle's speed;
- GPS to record the vehicle's location;
- an accurate time clock;
- a GSM mobile device to communicate location, speed and time information to a central location.

References

1. Institution of Highways and Transportation & Department of Transport (1987) *Roads and Traffic in Urban Areas*, HMSO, London.
2. Department of Transport (1981) *Traffic Surveys by Roadside Interview*, Traffic Advisory Leaflet 11/81, HMSO, London.
3. Greater London Council (1967–1986) *Greater London Speed Surveys*, Greater London Council.

3
Parking Surveys

3.1 Introduction

Every trip by a vehicle results in a parking act at the end of the trip. The importance of parking can perhaps be illustrated by the fact that, on average, a car in the UK is parked for about 23 hours a day. The vehicle may be parked on the street or off-street in a car/lorry/cycle park, or in a private garage. How vehicles arrive and depart from these parking places, how long they stay and under what circumstances define vehicular traffic and indeed some pedestrian traffic on the roads and help to determine what measures are required to meet or manage the demand. Therefore, it is very important to obtain an objective and unbiased understanding of this activity by properly constructed and conducted surveys.

To allow us to understand the parking behaviour, there are a multitude of parking survey techniques, which have been developed, each aimed at measuring something slightly different. Each technique is discussed below, in terms of the order of complexity.

3.2 Choosing when to survey

Activity will vary from day to day and season to season, and theoretical statisticians would no doubt expect any survey to be repeated for a representative number of days or periods in order to ensure a completely unbiased and representative sample, or to attach levels of uncertainty to the outcomes of surveys if collected on a single day. Unfortunately, traffic engineers seldom have the luxury of either the time or resources necessary to undertake multiple repeat surveys and have to make hard decisions based on the data they have. Therefore, we usually have to compromise with a survey on a single day and make decisions based on the results obtained, tempered by experience and common sense. In order to get the best value for money out of such a survey, it is important to take maximum advantage of any pre-existing knowledge or information.

Thus, for example, if one wishes to understand 'normal' peak demand in a shopping area one might have traffic flow data which showed the busiest day of the week. Typically, traffic activity increases slightly from Monday to Friday, but shopping activity might be greatest on a Saturday and local retailers might be able to provide guidance on this. Similarly, it is clear that sales periods and the pre-Christmas rush is very busy but abnormal in that the situation only occurs for a few days a year and unless one were seeking to plan for these exceptional events, such surveys would not provide an understanding of normal conditions.

We might also reasonably believe that shopping activity would be lower during school holiday periods when people are more likely to go on holiday or to spend time with their children on other activities, although, of course, the converse would be true in areas where there is a high level of tourist activity.

From the above we can see that we have begun to identify, from other information, a target time slot for a shopper parking survey which:

- excludes school holidays and periods of abnormal demand, such as Christmas;
- tends towards the end rather than the beginning of the week, probably favouring Friday or Saturday.

Thus we can begin to select a preferred time for our survey and with limited resources target one or two days which are more likely to give us a measure of what we are trying to understand.

3.3   Supply surveys

To understand existing parking behaviour, and the potential for accommodating additional parking, it is essential to have a sensible estimate of the amount of parking available in a given location. This is not always as simple as it may sound, since cars and other vehicles can be parked in many different places.

On the street, in controlled parking areas, street parking is explicitly marked either as individual parking bays or as lengths of street where parking is allowed. The bays can be counted explicitly; for lengths of road where parking is allowed, it is reasonable to use an estimate of 5 metres of kerb space for each car parking space. The figure of 5 metres is derived empirically from observations in many surveys. When undertaking a survey of the spaces available on-street, it is important to remember that restrictions may only apply part of the time. This means that the supply of available parking space could vary, according to the time of day or day of the week.

Off-street, land and structures which are designed to be used as parking are often marked out with car parking spaces, which can be counted explicitly. However, it is commonplace to see yards, service roads and other areas, which were never intended as formal parking, used for parking on a regular basis. These can make a significant contribution to total parking supply. For example, in central London, a 1977 Census survey,[1] where all the places regularly used for car parking were recorded, identified some 57 000 car parking spaces. A place was recorded if there was a vehicle parked or there was evidence, such as oil stains or exhaust marks which gave evidence of regular car parking. A later study, based on planning records,[1] suggested that there were about 34 000 spaces in the same area. This is a 40% difference which is largely explained by the many places regularly used to park cars which were not formally identified as being for that purpose. More recent experience from other surveys suggests that a 25% difference between 'formal' and 'actual' parking capacities might be typical.

Therefore, in order to understand parking behaviour the first step is to accurately measure the amount of parking available in the study area, and experience suggests that the only reliable way of doing this is to actually walk through the streets and count the spaces, as formal records can be a quite unreliable estimate of the true situation.

3.4   Occupancy surveys

The simplest parking activity survey is an occupancy survey, where the number of vehicles parked on a street, in a car park or parking area are periodically counted. A surveyor passes round the parking spaces at pre-determined intervals and simply counts the number of vehicles in the parking place. The surveyor may record:

- the total number of vehicles;
- the number of vehicles in each street or length of street;
- the numbers of each type of vehicle, by street or street length.

This survey technique tells us little about the vehicles, in terms of their arrival, departure and duration of stay; however, it does allow us to gauge the adequacy of the car parking available, when compared with parking demand. The surveys also tell us how busy the parking is at different times of the day or week.

This survey technique is appropriate where the data is being collected to give either a broad understanding of the adequacy of the parking supply or an understanding of changes in demand over time. Thus, for example, if a car park operator wished to ensure that he always had enough parking available to be sure that a driver could always expect to find a place to park, he might set a threshold of 85% occupancy and, when demand reached this level, he would either increase supply or make the parking less attractive, by, for example, raising charges, so that demand was kept below the threshold.

The technique allows a large amount of parking to be surveyed quickly and so requires fewer survey resources than the other, more precise, methods described below. It provides us with good information about gross levels of activity but tells us nothing about the behaviour of drivers.

3.5 Beat surveys

If we wish to have more details about the behaviour of individual vehicles, and hence an aggregate picture of parker behaviour, not just the gross level of parking demand, we can use a beat survey. In a beat survey the surveyor visits, in turn, a pre-determined number of parking spaces and records details of the vehicles that are observed parking in each space. Typically, the surveyor would record:

- time;
- parking space location, this is required to allow successive observations to be compared;
- vehicle type;
- partial vehicle registration number (described below).

Normally, a beat survey is undertaken at regular intervals and so the time is recorded to an appropriate time block. Thus if the survey were hourly, the time would be recorded as the hour in which the survey round took place, and so on. Figure 3.1 shows an example of a survey form for a street survey of a typical area.

The beat frequency will be determined by the purpose of the survey. If an area were used mostly by residents, who tend to park all day, or by workers who arrive in the morning and leave at the end of the working day, then a survey may only be required every 2 hours. However, if the survey were in a high street, where vehicles are coming and going every few minutes then a 15-minute beat might be more appropriate.

More commonly, however, the survey technique is used to understand patterns of arrival and departure and duration of stay within a single day or part of the day. It can be used to distinguish between all-day and short-stay parking activity.

If one wished to identify the number of long-stay parkers, three or four visits a day would allow an unambiguous quantification of long-stay parking; however, it could considerably under-count short-stay numbers. We return to discuss this issue in greater detail later.

Parking / loading survey

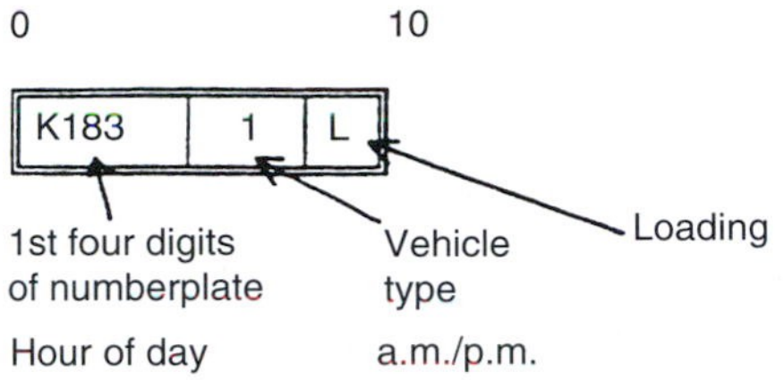

0   10

| K183 | 1 | L |

1st four digits   Vehicle   Loading
of numberplate   type

Hour of day   a.m./p.m.

Vehicle types

1	Motorcycle	5	Medium goods
2	Car/light van	6	Heavy goods
3	Medium van (transit)	7	PSV
4	Taxi	8	Other

L = Loading

Section A

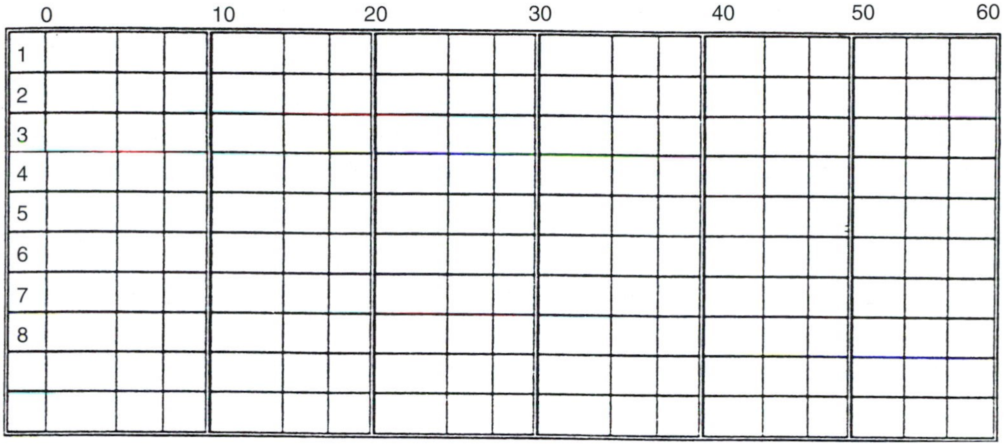

0   10   20   30   40   50   60

| 1 |
| 2 |
| 3 |
| 4 |
| 5 |
| 6 |
| 7 |
| 8 |

Section B

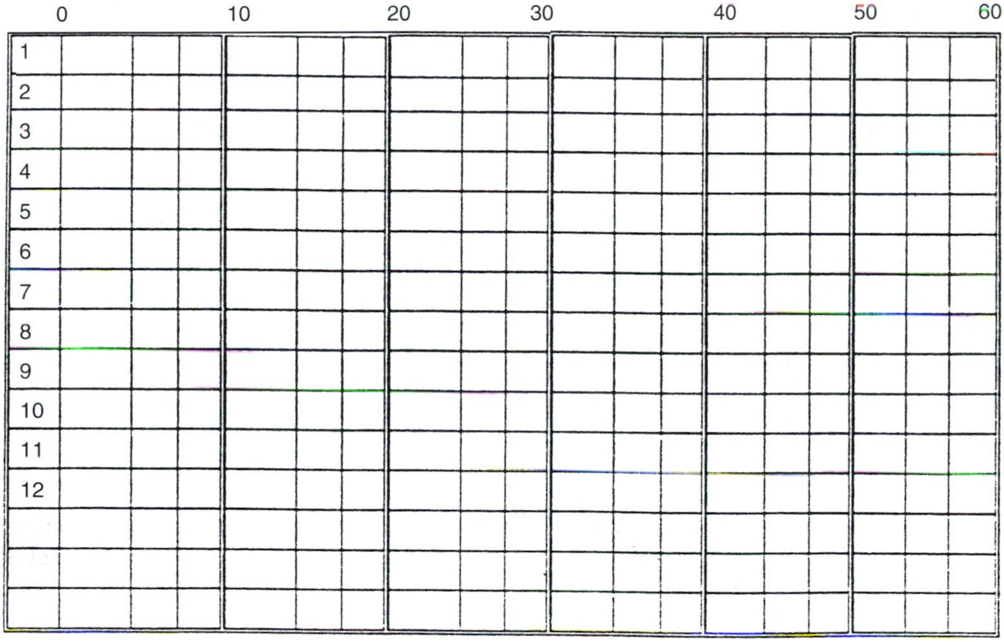

0   10   20   30   40   50   60

| 1 |
| 2 |
| 3 |
| 4 |
| 5 |
| 6 |
| 7 |
| 8 |
| 9 |
| 10 |
| 11 |
| 12 |

Fig. 3.1 Parking beat survey form.

Where it is important to identify short-stay parking, then the beat frequency needs to be much higher, possibly as often as four times an hour. Even this can under-count very short duration acts. The methods that can be adopted to address this problem are discussed later.

If data are to be compared between succeeding beats, to see if a vehicles has moved, it is very important that the parking act being observed can be located exactly to a particular parking space. If this is not done, we could have the situation where the act observed in space 'a' on one pass of the surveyor is compared with an act observed in space 'b' in the next round.

It is also very important to ensure that the surveyor passes round the beat in the same order at each visit, to avoid distortion of the results. The reason for this can be explained with a simple illustration. Figure 3.2 shows the path a surveyor is expected to follow round a car park for a 30-minute beat survey, starting at **S** and ending at **F**. If the survey takes 25 minutes to complete, with a 5-minute break at the end of each pass, then successive observations at each bay will be about 30 minutes apart. If, however, after completing a beat in the correct order the surveyor was to retrace his steps, visiting the parking bays in reverse order, then the two observations of the last bay would be only 5 minutes apart, while the two observations of the first bay would be separated by an hour.

Beat surveys can be used for surveys on the street or in a parking area. In either case careful planning is needed to ensure that each parking place can be recognised, to allow correct comparison of data between successive beats.

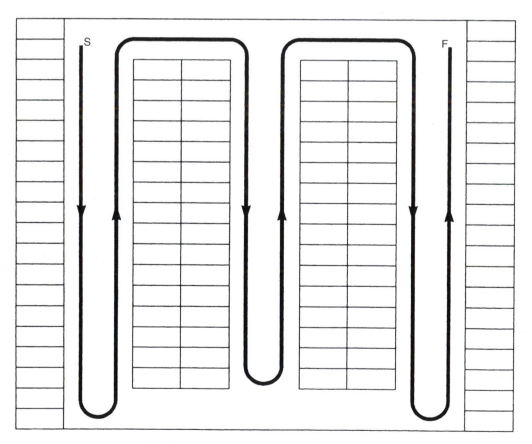

Fig. 3.2 Example of car park beat survey route.

In most car parks the bays are marked and so location can be explicitly identified, provided the same route round the car park is followed. In a street, where there may be no markings, location is less clear and can be confusing. To avoid this, the street should be divided into nominal parking bays at the planning stage of the survey (Figure 3.3). On the day of the survey the bay limits could be marked on the kerb with chalk, or some other non-permanent marker, to aid the surveyors.

Obviously in a street with no markings drivers will not neatly park in the nominal survey bays. Any parked vehicle should be allocated to the bay it most nearly sits in. This should not cause confusion as, if a vehicle straddles two bays, it should be allocated to the bay it occupies most of. Clearly until the first vehicle leaves, a second vehicle cannot occupy the same space, so no confusion should occur.

If the purpose of the survey is to understand both the type and number of vehicles parked, then the type of vehicle seen should be recorded. Even if this information is not required, it can still be a useful way of checking data quality and helping to clarify uncertainties about the survey results.

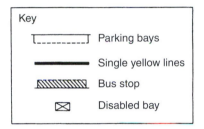

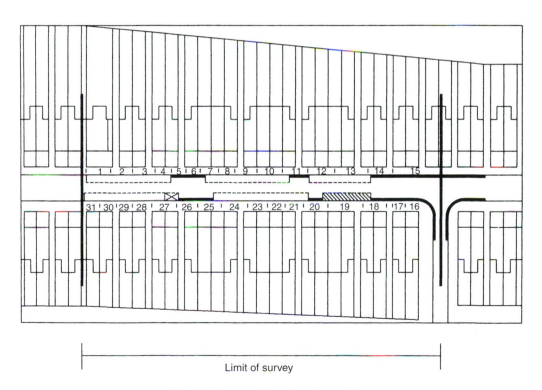

Fig. 3.3 Survey design for street parking.

The complexity of vehicle types recorded will depend on the use to be made of the survey results. A minimum might be 'cars' and 'other vehicles'; however, if one wishes to gain a more detailed understanding of activity, 'cars' could be sub-divided into car, taxi and blue badge holder, and 'other' might be split into motorcycle, buses and various sub-categories of commercial vehicle. Figure 2.7 shows the standard vehicle classification used by the Department for Transport, and if this is used it will allow a consistency between surveys.

Whatever the categorisation adopted, it is clearly important to ensure that surveyors are fully and unambiguously briefed on how to identify the types of vehicle to be used for the particular survey.

Such a survey could also be used to identify vehicles that regularly park in the same place, for example commuters who park in the same street or streets every day. In this case the survey could take place just once a day, over say a week. One would be trying to track vehicles that would move location from day to day, but stay in the same general area. The exact location would not be so important but it would be necessary to record the full registration mark so that the comparison can be made reliably.

Beat surveys compare a series of snapshots of activity and we can explain the parking behaviour of the vehicles in the area being studied by identifying the change between the snapshots. However, because the survey only observes the vehicles intermittently, we do not have a completely accurate picture of the behaviour we are observing. The survey technique suffers from two main types of inaccuracy. These are described below.

3.5.1 TIMING ACCURACY

When a vehicle is first seen on a beat the observer does not know exactly when it arrived, only that it arrived after his last visit and before the present one. Similarly, when the vehicle leaves the observer does not know the exact time of departure. It can be seen that if the beat frequency is once every t minutes then, at one extreme, the vehicle could have been parked for up to $2t$ minutes longer than the observer has logged. At the other extreme the vehicle could have arrived just as the observer reached the parking place on one pass and left just as he/she left the parking place on another pass, in which case there would have been a zero error on the time recorded. It follows that the average error of observation is t, the beat frequency, and so when calculating length of stay all observations should have this amount added to give an unbiased estimate of duration of stay, that is if a vehicle is recorded n times, then the best estimate of its duration of stay is nt minutes.

3.5.2 UNDER-COUNTING

With a beat survey a certain number of vehicles will arrive and depart between successive passes of the surveyor without being seen. Thus short-stay parking acts, that is those with a duration of stay less that the beat frequency, will always be under-counted on a beat survey. The scale of under-counting and its importance will depend upon the specification and purpose of the survey.

For example, in a long-stay car park, where the purpose of the survey is to measure the amount of long-stay parking, the fact that short-stay acts have been under-represented will be of relatively little importance. However, in a survey aimed at recording short-stay illegal parking in a restricted street, a beat could significantly under-count the number of such acts.

This factor was first recognised explicitly as a potential major error in the mid-1970s when it became clear that central London suffered from extensive illegal parking and yet standard beat surveys were failing to satisfactorily explain this behaviour. In 1977 the Greater London Council undertook a comprehensive survey of parking in central London.[1] As part of this study the

Table 3.1 Level of under-counting with a 30-minute beat survey

Parking type	Acts observed (%)
Two hour metres	89
Four hour metres	93
Residents' bays	90
Single yellow line	41

(Source: Adapted from Ref. 1).

vehicles observed in a 30-minute beat survey were compared with the actual number of vehicles parking recorded by continuously observing the same streets. The results are shown in Table 3.1.

Table 3.1 clearly shows that, although the under-counting on the longer-stay (compared with the beat frequency) parking is around 10%, for the short-stay (illegal) parking the survey only recorded about two in every five acts. Clearly, if the beat data had been used as a basis for judging the level of illegal parking, the scale of the problem would have been massively understated and any conclusions based on the results would have been invalid.

Recognition of this deficiency led to the development of the continuous observation street survey technique described below.

3.6   Continuous observation survey techniques

If we do not need to know where within a larger car park a particular car has parked, it may be more efficient to simply record the vehicles as they enter and leave the car park. In this case, the time of arrival, vehicle type and registration are recorded on entry and again on exit from the car park. The two data sets are then compared to establish how long each vehicle was parked.

This technique is more accurate than a beat survey as the duration of stay is known exactly, within the accuracy with which the data is recorded. With this type of survey, often a large car park can be observed with just two or three people. This task can also be undertaken by using video cameras to observe the entry and exit points, with analysis carried out in the office.

The data can be recorded on a form of the type shown in Figure 3.4. Alternatively at a very busy car park, the surveyor could record the data using a tape recorder, for later transcription and analysis. However, tape recorders should be used with care, as they introduce a whole different set of risks such as, flat batteries, reaching the end of the tape at an inopportune moment, or being unable to hear the surveyor over background noise.

The accuracy of an entry/exit survey in a car park can be transferred to observing parking in the street with a continuous observation survey. With this type of survey, each surveyor is limited to watching just a few parking places, as many as can be seen from one location. Typically, this would be 30–40 spaces.

The surveyor is able to exactly record the arrival and departure time of each vehicle and, if appropriate, can also record other relevant information, such as:

- whether or not the vehicle was loading;
- whether the vehicle was illegally parked;
- if the parking was paid for and, if so, when the payment expired.

The surveyor can also record details, such as enforcement activity, if this is required. Although such continuous observation techniques should provide a very detailed understanding of parking

Car park name................................   **Site number**.......   **BUTP II PARK3**

Surveyor's name..................   **Date**..........

Weather: dry/rain

Vehicle registration	Type	Time in	Time out		Vehicle registration	Type	Time in	Time out

Notes:

Vehicle type: 1 Car
2 Taxi
3 Two wheeler
4 Other

Fig. 3.4 Example of continuous survey form.

behaviour, albeit in a relatively small area, one must recognise that the presence of a large number of survey staff with clipboards and wearing high visibility jackets could have an impact on driver behaviour, particularly those who were intending to park illegally. To overcome this, one might use remote observation using a strategically positioned video camera to observe parking behaviour in a less conspicuous way.

In 1982, the Transport Research Laboratory (TRL) developed a computer package specifically designed to undertake this type of survey called PARC.[2,3] However, this package is no longer available, and so anyone wishing to undertake such surveys would have to develop their own analysis system.

In the parking activity survey techniques discussed above, there is generally a trade-off between coverage and depth of detail. Thus, with a very simple occupancy survey, a surveyor can cover a very large area in a given time period whereas, at the other extreme, for a continuous observation survey, using the same resources, the surveyor can only cover the spaces that can be seen from a fixed location.

3.6.1   *RECORDING NUMBERPLATES*

The discussion above suggests that in most cases time can be saved by recording a partial numberplate, on the assumption that the chances of two vehicles having the same partial index and being in the same survey area is sufficiently small not to be a concern. UK numberplates have made several evolutions over the years viz:

> ABC 123 prior to 1963
> ABC 123 D 1963 to 1984
> A 123 BCD from 1984 to 2001
> AB 12 BCD from 2001

One should note, however, that with the latest evolution of the UK numberplate system the first two characters represent the geographical area in which the vehicle is registered and the two numbers represent the half-year. Therefore, if one were to record the first four characters there would be a significantly higher chance of having two vehicles with the same first four characters than if one were to record the last four.

3.7   Summary

To summarise, the engineer wishing to survey parking supply or activity has a wide range of techniques to call on. The method chosen will depend upon the rationale for the survey. It is important to understand which survey is appropriate for the information required and to ensure that the correct technique is selected.

References

1. Carr, R, Baker, LLH and Potter, HS (1979) *The Central London Parking and Car Usage Survey*, Greater London Council.
2. JMP Consultants Limited (1989) *PARC Suite Users Manual*, JMP, London.
3. Transport and Road Research Laboratory (1991) *Guide to PARC.2P*, HMSO, London.

4

Estimating Travel Demand

4.1 Introduction

The estimation of travel demand is a fundamental part of traffic engineering design work. The key questions are how much effort needs to be expended in estimating demand and what method should be adopted. The answers depend on the nature of the design issues. For example, a minor traffic management design to improve road safety over a length of road in inner London where traffic flows have been stable for many years will require little more than a survey of existing traffic. The reverse is true of a proposal for a new roadway to assist regeneration in an old urban area where design will depend on estimating the new traffic likely to be attracted to use the new road.

Estimation techniques fall into three main categories:

- growth factor
- low-cost manual estimation
- computer-based traffic models.

All of these techniques include assumptions about the four basic elements of estimation, which are:

- trip generation
- trip distribution
- modal split
- assignment.

A 'trip' is defined as a one-way travel journey between the origin (start) and destination (end) of the journey. Trip generation is the number of trips starting or ending at an area (or zone) in a given time period, for example day or hour.

Trip distribution describes the number or proportion of trips from an origin zone spread among all destination zones.

Modal split is the split (or share) of these trips among different modes of travel (e.g. car, public transport, walk, cycle).

Assignment is the process whereby trips are routed from their origins to their destinations through a travel network (e.g. road, bus and train routes, cycle ways/routes).

These four basic elements, shown in Figure 4.1, define the numbers of trips made from an area, the destination of these trips, the modes of travel used and the routes taken.

The choice of estimation technique will depend on the complexity of the demands to be estimated and the resources available for the estimation. For example, if an estimate of the future

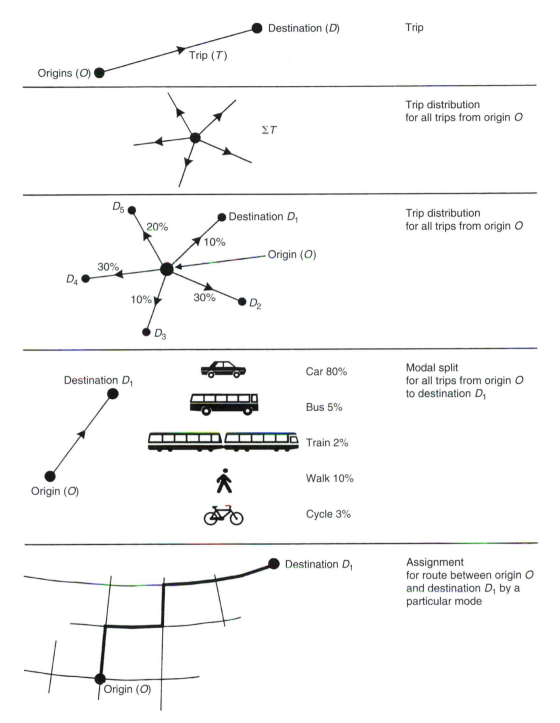

Fig. 4.1 Basic elements of a trip.

number of vehicles on a rural road is required, then a growth factor method may be appropriate. If the impact of a new small housing development on the local road network is to be assessed then low-cost manual estimation is likely to be adequate. Estimation of travel demands by different modes of transport in an urban area will probably require the use of computer-based traffic models.

4.2 Growth factor

The Department of Transport has prepared National Road Traffic Forecasts[7] (NRTF) which are based on national models with two alternative sets of assumptions on car ownership and usage and on goods haulage. The resultant two sets of forecasts of traffic growth are known as 'high growth' and 'low growth'. These forecasts are of increased vehicle kilometrage, which translates directly into an increase in traffic on a section of road.

The forecasts have been prepared to year 2026 as shown in Table 4.1. As an example, the forecasts predict a growth in traffic between 1996 and 2011 of between 27% (low growth) and 43% (high growth).

Use of these forecasts is appropriate only when counts of traffic are available for the particular section of road being studied. When the origins and destinations of trips on the road have been surveyed (or are available from a traffic model) then local growth factors rather than the national forecasts can be used. These local growth factors are also based on the national models and are available at local authority district level in these models. At this disaggregated level, they reflect local zonal projections of planning data. The local factors are available as trip-end estimates for each District from the Department of Transport's TEMPRO program. A national forecast adjustment factor (NFAF) then has to be applied to make these local forecasts of travel compatible with the NRTF, which records the forecast growth in vehicle kilometres. In its simplest form, this approach can give an estimate of future growth as the average of the local factors weighted by the trips from each origin and then adjusted by NFAF.

This growth factor approach then estimates the future trip generation and distribution while assuming that modal split and assignment remain constant.

Table 4.1 Traffic growth

	Year	Low growth index for all vehicles	High growth index for all vehicles
Observed years	1976	1.000	1.000
	1979	1.052	1.052
	1984	1.244	1.244
	1989	1.691	1.691
	1994	1.761	1.761
	1996	1.827	1.867
Forecast years	2001	1.989	2.136
	2006	2.153	2.402
	2011	2.316	2.666
	2016	2.483	2.924
	2021	2.648	3.184

4.2.1 INTRODUCTION TO TEMPRO

TEMPRO, the trip-end model presentation program, is designed to allow detailed analysis of pre-processed trip-end, journey mileage, car ownership and population/workforce planning data and present this either on screen or to file/printer for distribution via hardcopy. The pre-processed data is itself the output from a series of models developed and run by DfT.

TEMPRO has been in existence for several years and prior to the current version dealt with motor vehicles only. The new version (Version 4) is now multi-modal, providing data on trips on foot, by bicycle, motor vehicle (both as a driver and passenger) by rail and by bus.

Analysis of the data can be any combination of:

- geographical area (down to a 'zonal' level of resolution, with each local authority area divided into several zones);
- transport mode;
- time of day of travel;
- purpose of journey;
- years of interest (from 1991 to 2031);
- type of analysis (trips in terms of producing/attracting, origin/destination or car ownership).

Results are presented either in terms of growth over the selected period or in terms of the raw trip-end data.

The model also allows the underlying planning data used in generating the pre-processed data to be viewed and alternative planning assumptions to be entered to see what impact these would have on trip-end data.

4.3 Low-cost manual estimation

The starting point for low-cost manual estimation will normally be an origin–destination (O–D) matrix of traffic. This matrix could be obtained from a survey of traffic using a section of road, normally by roadside interview, and this would be appropriate if a local improvement to part of the road network is proposed.

Alternatively, if expansion of existing developments or new developments is planned, then a survey of existing residents, employees and visitors or of residents, employees and visitors at nearby similar developments can provide the O–D information required.

The matrix will be growthed to the forecast future year by using trip-end growth factors, that is factors applied to the origins and destinations of the matrix. These factors can be the local growth factors by district as previously discussed, locally calculated factors at a more disaggregate level, or values forecast for new developments as described in Chapter 14. There are several simple matrix-manipulation techniques such as the Furness technique for applying these growth factors to the O–D matrix.

The Furness technique involves an iterative balancing of the rows and columns of the matrix so that the growth in origin and destination trip ends, as well as the new total flow, represented by the total of the matrix elements, are correctly represented as follows:

For

T_{ij} trips between origin zone i and destination zone j surveyed
O_i trip end for origin i surveyed
D_j trip end for destination j surveyed

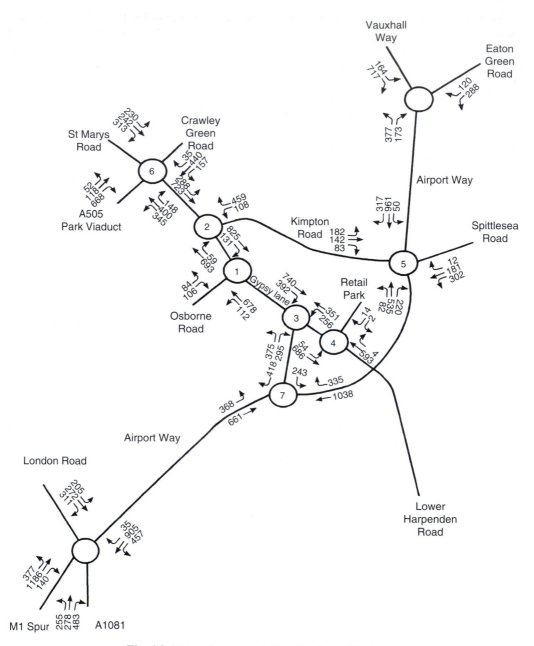

Fig. 4.2 Network representation showing traffic flows.

and NFAF-adjusted growth factors

g_i for origins
g_j for destinations

then

new origin trip end for zone i $= O_i g_i$
new destination trip end for zone $j = D_j g_j$
new total flow $= \sum O_i g_i$

The next step is to assign the matrix to the road network. In its simplest form of representation, the network will consist of road links with measured or estimated journey times and lengths. A simple network representation is shown in Figure 4.2. The assignment process typically involves identifying the quickest route from each origin to each destination based on journey time or the shortest route based on distance.

Traffic between origins and destinations is then loaded onto the network; for a small network (e.g. up to 50 links) and a small matrix (e.g. up to some 50 elements) then manual loading may be practical. For more links and elements it will be necessary to consider the use of automatic network assignment techniques described in the next section.

A more sophisticated approach than manual assignment is to use a diversion curve[2] and this may be appropriate when two competing routes are available, for example when a local bypass is being considered. The approach allows for multiple routeing to represent the different perceptions of best route that drivers have, and provides a more accurate estimate of the proportion of existing traffic that will switch to using a bypass.

When considering new or expanded developments, the forecasting is undertaken in two parts. The traffic from the new developments is estimated based on O–D data and assigned to the network. Existing traffic on each link is then growthed to represent future years using the factors described and then added to the traffic from the new developments. This is described in more detail in Chapter 14.

The low-cost manual estimation then adds a manual assignment to trip generation and distribution but assumes that modal split remains constant.

4.4   Computer-based traffic models

A traffic model is a set of mathematical equations which, when taken together, provide an estimation of traffic flows. Some models are designed to investigate strategic planning issues at a broad level, to assess specific policy options, such as congestion charging, or to identify the interaction between land-use changes and transport. These models are unlikely to be used directly by traffic engineers and so this section deals only with serial models, such as SATURN[5] and CUBE (TRIPS)[6] which provide the four basic elements of estimation:

- trip generation
- trip distribution
- modal split
- assignment.

Typically the model will be disaggregated by time of day, for example AM peak, inter peak and PM peak. The trip generation or trip-end model may also be disaggregated by purpose and vehicle type, for example:

- person home-based work trips;
- person home-based education trips;
- person home-based shopping trips;
- person home-based other (e.g. leisure and social, trips);
- person non-home based (e.g. employer's business, trips);
- commercial vehicle trips;
- other vehicle trips (e.g. bus, coach and taxi).

The trip-end model will be based on relationships involving population, employment, car ownership and land-usage characteristics and often determined using linear regression.

Trip distribution is often based on a model employing the concept of generalised cost of travel. Generalised cost (C) puts a monetary value on the distance travelled and the time spent travelling and includes other travel costs, so that:

$$C = a \times \text{travel time} + b \times \text{travel distance} + d$$

where

      a   is the value of time as assessed by the Department of Transport,[4]
      b   is the driver's perceived vehicle operating cost,[4]
      d   is other travel costs, for example parking and toll charges.

The common form of the model is:

$$T_{ij} = O_i D_j f(C_{ij})$$

where T_{ij}, O_i and D_j are as in Section 4.3 and

$$f(C_{ij}) = \alpha e(-\beta C_{ij})$$

with α and β being estimated parameters.

Typically α enables the resultant matrix to balance to correctly represent total traffic flow and β is calibrated against surveyed flows. Where calibration is not possible, then the deterrence function is often simplified by traffic engineers to:

$$f(C_{ij}) = \frac{\alpha}{C_{ij}^2}$$

which is comparable with Newton's gravity model.

In urban areas, it is now common to assess the modal share of travel, that is the proportion of travel by private transport, public transport, walk and cycle. The logit model is often used and is of the general form:

$$T_{ijk} = \frac{T_{ij} e(-\lambda_k C_{ijk} + \delta_k)}{\Sigma_n e(-\lambda_n C_{ijn} + \delta_n)}$$

where

    T_{ijk} is the number of trips between zones *i* and *j* by mode *k* (e.g. public transport);
    T_{ij} is the number of trips between zones *i* and *j* by all modes;
    n   is the number of modes being considered;
    λ_k, λ_n is a cost parameter for mode *k* or *n*;
    C_{ijk}, C_{ijn} is the generalised cost of travel by mode *k* or *n*;
    δ_k, δ_n is the modal handicap, for example the monetary value travellers place on being able to use private transport rather than public transport.

Trip assignment models in basic form identify the least-cost route through a transport network between origin and destination and then load the estimated O–D matrix for a travel mode onto that network. This basic form is known as 'all or nothing' as all traffic is assumed to select only one route for a journey between an origin and a destination.

Two refinements are commonly adopted; these are stochastic assignment and capacity restraint assignment. Stochastic assignment takes account of the different perceptions that drivers have of the best route and so spreads traffic over more than one route. Capacity restraint assignment allows for the increase in travel time caused by congestion and the resultant decision by drivers to use a different route to their preferred route. This is often achieved using an equilibrium approach which has as its basis that traffic on a congested network arranges itself such that all routes used between an origin and a destination have equal minimum costs.

A standard procedure in developing a traffic model is to first calibrate and validate it before it can be used for forecasting travel demand. Calibration and validation requires the development of observed trip O–D matrices.

Ideally, these matrices are based on surveyed travel data. However it is not possible to obtain a full coverage of an urban area and so two techniques are commonly used; partial matrix and matrix estimation. The partial matrix technique involves 'filling-in' missing parts of the matrix using the travel data that is known. The matrix estimation technique involves the use of the method of maximum likelihood to estimate the matrix using road link counts; this technique can also be used to update a matrix obtained in the past.

Once the model parameters have been established by calibration against the observed O–D matrices, then an independent set of traffic data, typically counts on screenlines, is used to 'validate' the accuracy of the model. The model is then ready for the traffic engineer to forecast future travel demand for a set of planning assumptions on future land-use location and transport infrastructure.

4.5 Micro-simulation

Micro-simulation models use sophisticated computer techniques to model the behaviour of individual vehicles and their drivers. This approach leads to a closer representation of activity on the highway network than is possible using traditional average modelling techniques such as those employed by program suites such as SATURN, TRIPS, EMME/2, etc.

A key feature of micro-simulation models is the graphical representation of the network showing the movement of the individual vehicles in real time, whether on a simple representation of the network or against a mapping background, as shown below.

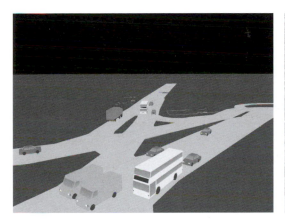

Due to the 'real-time' simulation methodology, it is possible to simulate and assess traffic behaviour over a wide range of conditions, from free-flow highway intersections to traffic signal-controlled junctions in congested urban areas. Sophisticated traffic signal systems, including vehicle actuation and pedestrian facilities can be modelled, including the behaviour of SCOOT regions. Micro-simulation is particularly suitable for assessing alternative traffic management strategies, as the interactions between the various traffic streams are immediately obvious from the screen displays.

For public exhibitions, or other situations where there is a need to communicate the results of transport studies to a wide audience, the ability to display the results of the forecasts in this way allows the debate to focus on the merits of the scheme(s) rather than the theory behind the modelling.

Micro-simulation packages currently available in the UK and mainland Europe, include:

- PARAMICS
- VISSIM
- AIMSUM
- DYNASIM.

One example of the use of micro-simulation was to carry out an assessment of a proposed development on the heavily congested south-western quadrant of the M25. The Highways Agency was concerned with the impact on the main carriageway flow due to the increased weaving downstream of the proposed site. Micro-simulation allowed the collection of detailed information on the number of lane changes, the individual journey times of each vehicle passing through the network, and a count of urgent manoeuvres to avoid collisions.

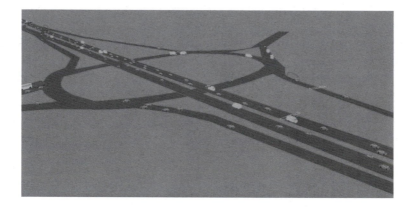

Congested road networks can experience a rapid transition into unstable flow, with shock waves being sent back through the following traffic. This situation is highly dynamic and depends on the interaction between specific vehicles. This sort of mechanism does not lend itself to analysis by traditional steady state macroscopic models, and so the use of micro-simulation allowed analysis that would not have been possible previously.

4.6   Accurate and appropriate data

However good the technique chosen for estimating travel demand, its value depends on the accuracy and appropriateness of the traffic data used. The main issues in determining the amount and type of data to be collected are the time period for appraisal and the purpose of the demand estimation.

The Department for Transport (DfT)[3] requires that daily traffic levels are established for trunk road appraisal. This means that any roadside interview surveys should cover at least 12 hours of the day to provide a reasonably accurate daily O–D matrix. Automatic traffic counts are typically used to establish annual average daily traffic (AADT) levels and to provide the factor to convert 12-hour survey information to represent 24 hours.

The DfT[1] also requires that junction merge and diverge and weaving section calculations be based on a percentile highest hour flow. Roads are classified into three types, based on a seasonal

Table 4.2 Trunk road types

Road type	Seasonal coefficient	Percentile highest hour flow
Main urban	1.05<	30th
Inter-urban	1.05–1.25	50th
Recreational/inter-urban	>1.25	200th

coefficient with the percentile highest hour flow to be estimated for demand and appraisal purposes shown in Table 4.2. Automatic traffic counts can provide the percentile highest hour flow.

$$\text{The seasonal coefficient} = \frac{\text{flow in August}}{\text{flow in neutral month, usually May}}$$

Neutral months are taken as April, May, June, September and October, and it is these months that are normally selected for surveys on trunk roads and also non-trunk (county) rural roads.

Peak period analyses are normally required for non-trunk urban roads. Surveys must then cover the peak periods and identify the profile of flow within these periods. Typically, a highest flow over 1 hour, known as the peak hour which can vary from site to site, is identified and used for analysis. Taking a turning movement count at an urban junction over a single day will give a measure of the highest flow. Its accuracy as representative of average peak hour conditions over the year will be relatively low because of variations of flow from day to day. This accuracy can be enhanced by adjusting the single-day count to represent average flows obtained from automatic traffic counts over a longer period. Any attempt to assess average peak hour conditions must allow for this variation in flow that occurs naturally.

Individual counts are themselves subject to errors that are not related to this variation. These errors can occur from three sources:

- data sampling errors
- measurement errors
- human errors.

Sampling errors typically occur in surveys. For example, a roadside interview survey will often only achieve an interview sample of some 25% of drivers passing the interview point; these 25% may be unrepresentative of the total. The value of the sampling error can be estimated if reasonable assumptions are made about the way in which the sample is drawn.

Measurement errors depend on the method of measurement. Studies have shown that an automatic traffic count of 1 day will be accurate to within ±5% of the actual count to a 95% confidence interval. Manual classified counts give accuracies of within ±10% for all vehicles and ±18% for all goods vehicles, again within a 95% confidence interval.

Human errors occur when data is processed and analysed. Quality checking procedures are needed to minimise them.

The forecasting process itself introduces further errors as any predictive model relies on assumptions that cannot be proven in advance. Estimates of travel demand then must always be considered as only that, i.e. **estimates**, and any analyses undertaken on the basis of estimated travel demand should be subject to sensitivity tests. These tests should cover a range around the estimated demand to ensure that any traffic engineering decisions on implementing physical changes or introducing new traffic regulations are robust against inaccuracies in the estimated demand.

References

1. Department for Transport (1996) *Design Manual for Roads and Bridges*, Volume 13 Section 2, Highway Economics Note (HEN) 2, HMSO, London.
2. Leeds, ITS and Atkins, WS (1995) *SATURN 9.2 User Manual*, Leeds.
3. Department for Transport (1996) *Design Manual for Roads and Bridges*, Volume 13 Section 1, Economic Assessment of Roads Schemes (COBA), HMSO, London.
4. Department for Transport (1996) *Design Manual for Roads and Bridges*, Volume 12, Traffic Appraisal of Roads Schemes, HMSO, London.
5. MVA (2003) *CUBE Manual*, Version 9.1, MVA, Woking.
6. Department for Transport (1992) *Design Manual for Roads and Bridges*, Volume 6, HA 22/92 Layout of Grade Separated Junctions, HMSO, London.
7. National Road Traffic Forecasts, 1989, Rebased at 1994, HMSO.

5

Capacity Analyses

5.1 Capacity definition

The term capacity when referring to a highway link or junction is its ability to carry, accommodate or handle traffic flow. Traditionally, capacity has been expressed in numbers of vehicles or passenger car units (PCU). (Vehicles vary in their performance and the amount of road space they occupy. The basic unit is the passenger car and other vehicles are counted as their PCU equivalent, e.g. a bus might be 3PCUs and a pedal cycle 0.1PCU.) In recent years public transport operators have applied pressure to consider highways in terms of their passenger-handling capacity and thus give a greater emphasis to the benefits of using high occupancy vehicles, such as buses or trams.

There is no absolute capacity value that can be applied to a given highway link, traffic lane or junction. Maximum traffic-handling capacity of a highway depends upon many factors including:

- The highway layout including its width, vertical and horizontal alignment the frontage land uses, frequency of junctions and accesses and pedestrian crossings.
- Quality of the road surface, clarity of road marking, signing and maintenance.
- Proportions of each vehicle type in the traffic flow and their general levels of design, performance and maintenance.
- The numbers and speed of vehicles and the numbers of other road users, such as cyclists and pedestrians.
- Ambient conditions including time of day, weather and visibility.
- Road user levels of training and competence.

The capacity of a road junction is dependent upon many of the features that govern link capacity with the addition of the junction type, control method and vehicle turning proportions.

The expression 'level of service' when applied to a highway refers to the *Highway Capacity Manual*[1] approach which defines a range of levels from the lowest which occurs during heavy congestion to the highest where vehicles can travel safely at their maximum legal speed (see Table 5.1; also DMRB TA 46/97[2]).

5.2 Effect of width on capacity

The capacity of a traffic lane is, within limits, proportional to its width. Clearly, there is a lower limit to the width of a lane below which it is operationally impractical to run vehicles. Below a lane width of about 2.0 metres capacity deteriorates rapidly. As lane widths approach the point

Table 5.1 Maximum service flow rates

Design speed (mph)	Level of service	Density (PCU/mi/ln)	Speed (mph)	Maximum service flow rate
70	A	≤12	≥60	700
	B	≤20	≥57	1100
	C	≤30	≥54	1550
	D	≤42	≥46	1850
	E	≤67	≥30	2000
	F	>67	<30	Unstable
60	B	≤20	≥50	1000
	C	≤30	≥47	1400
	D	≤42	≥42	1700
	E	≤67	≥30	2000
	F	>67	<30	Unstable
50	C	≤30	≥43	1300
	D	≤42	≥40	1600
	E	≤67	≥28	1900
	F	>67	<28	Unstable

where two narrow lanes can be marked or vehicles tend to form up in two lanes there is a rapid increase in capacity. In urban areas and at road junctions traffic will tend to form up in two lanes when the lane width exceeds 5.0 metres. The effect of lane width and saturation flows at traffic signals was demonstrated by Kimber *et al.*[3] Figure 5.1 shows the simplified relationship between lane widths and saturation flow.

5.3   Effect of gradient

A steep uphill gradient can significantly affect the acceleration rate of all vehicles when pulling away from stationary at road junctions. Heavy vehicle speed also deteriorates on a combination of gradient and length of gradient. Additional crawler lanes are provided on long steep gradients on motorways and other heavily trafficked roads to maintain speeds and capacity. Webster and Cobbe[4] recognised this factor in their work on traffic signal saturation flows and further work in TRRL RR67.

5.4   Effect of alignment

A tightly curving alignment in rural areas can cause a reduction in free-flow speeds. On existing roads tight curves are often accompanied by poor sight lines and forward visibility that prevents slow moving vehicles from being overtaken and reduces overall capacity. Bunching of vehicles, with reduced headways, can cause excessive delays at side road junctions.

In urban areas, curvature has been used to contain speeds in new residential areas and artificial curves, chicanes and horizontal deflection are used to reduce vehicle speeds as part of traffic calming schemes. This is discussed in Chapter 12.

5.5   Design flows

Design traffic flow is an arbitrary value chosen by the highway authority (HA) to reflect the highway capacity and prevailing local conditions, and includes a number of parameters, such as the

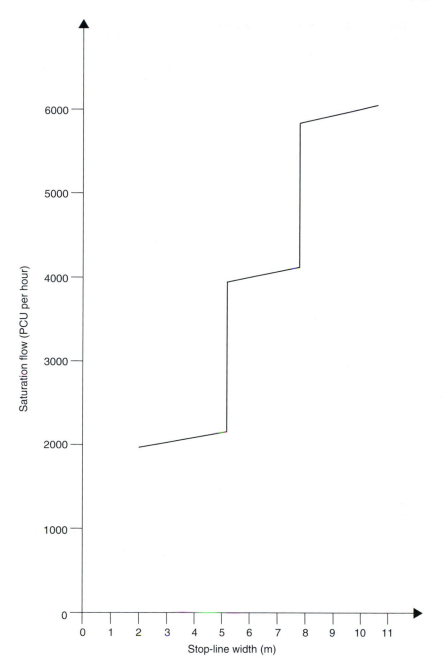

Fig. 5.1 Relationship between saturation flow and full stop-line width.

acceptable journey speed, free-flow conditions, acceptable delay, environmental impact, minimum levels of safety for vehicular and non-vehicular traffic.

The HA will consider the existing flows on a route, expected traffic growth, traffic generation from changes in land use and desirable modal split. Where an existing highway is to be improved, the potential to improve the route and its junctions will affect the decisions on design flow.

The design flows generally apply to a defined design year and usually refer to a minimum traffic-handling capacity for the improved route. Increasingly, HAs are considering maximum capacity and insisting that sustainable methods of transport are provided to limit the numbers of vehicle trips. Restraint on the numbers of car parking spaces at a development and subsidised public transport services are now an integral part of the design flow selection process. The success of the congestion charging in London has demonstrated that licensing the use of road space can have a positive effect on mode choice and reduction in peak traffic flows.

A single minimum figure for design flow is now rarely sufficient to define the design parameters for a highway scheme. Usually, a designer will be expected to provide space and capacity for alternatives to private modes of transport. Department of the Environment Guidance Notes PPG 6[5] and PPG13[6] emphasise the move towards sustainable development and transport.

5.6   Flow–capacity relationships

A measure of the performance of a highway or junction is the ratio of demand flow to capacity (RFC) or traffic intensity. As the RFC approaches 1.0 the level of congestion and queuing will increase. There are two basic opinions on the effects of traffic intensity on queue lengths: the steady state theory suggests that as traffic intensity approaches 1.0, queue length will approach

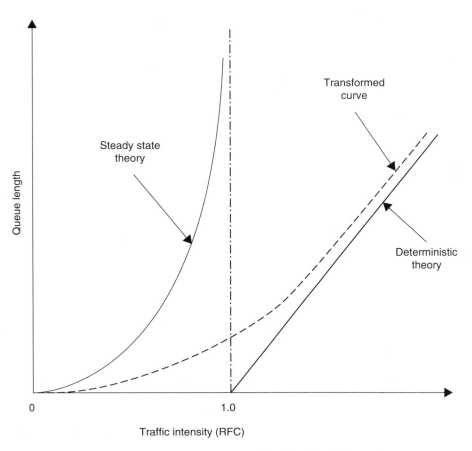

Fig. 5.2 Relationship between queue length and traffic intensity.

infinity; and the deterministic theory assumes that no queuing occurs until the RFC exceeds 1.0. In practice it can be easily observed that queuing starts to occur well before an RFC of 1.0 is reached and it is equally clear that queue length does not approach infinity at this point. The Transport Research Laboratory (formerly Transport and Road Research Laboratory), in its junction modelling computer programs ARCADY[7] and PICADY[8], has used a transformed curve for time-dependent queuing theory (Figure 5.2). This curve approximates to observable queues.

$$\text{Density } (D) = \frac{\text{Average number of vehicles in a length of highway } (L)}{L}$$

When density is zero flow is also zero; when density increases to a maximum there is no flow. Maximum flow occurs at some point between these values.

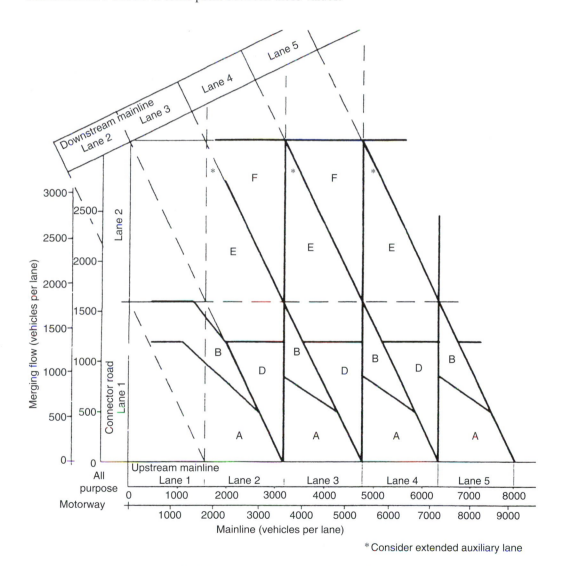

Fig. 5.3 Merging diagram for flow regions.

5.7   Junction capacity

The traffic movements at uncontrolled road junctions fall into three types of conflict: merges, diverges and crossing movements. At merges, two traffic streams travelling in approximately the same direction join together and combine into a single traffic stream. The capacity of the merge is determined by the capacity of the two upstream carriageways, the capacity of the downstream carriageway, the traffic intensity and the relative speed of each traffic stream. At diverges, a single

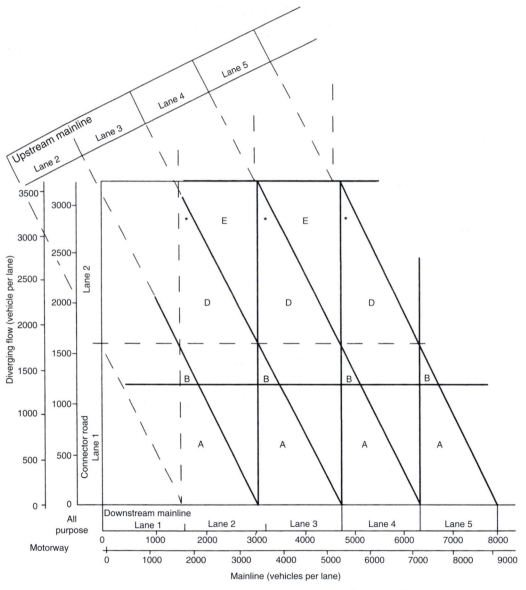

* Consider extended auxiliary lane

Fig. 5.4 Diverging diagram for flow regions.

Merge with no lane gain

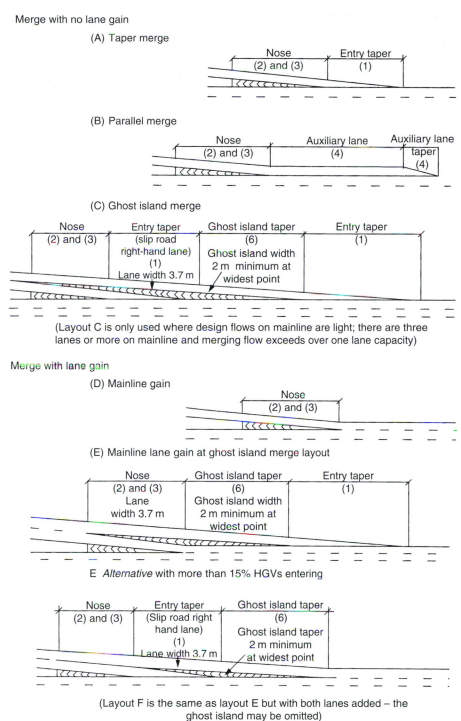

(A) Taper merge

Nose (2) and (3)   Entry taper (1)

(B) Parallel merge

Nose (2) and (3)   Auxiliary lane (4)   Auxiliary lane taper (4)

(C) Ghost island merge

Nose (2) and (3)   Entry taper (slip road right-hand lane) (1) Lane width 3.7 m   Ghost island taper (6) Ghost island width 2 m minimum at widest point   Entry taper (1)

(Layout C is only used where design flows on mainline are light; there are three lanes or more on mainline and merging flow exceeds over one lane capacity)

Merge with lane gain

(D) Mainline gain

Nose (2) and (3)

(E) Mainline lane gain at ghost island merge layout

Nose (2) and (3) Lane width 3.7 m   Ghost island taper (6) Ghost island width 2 m minimum at widest point   Entry taper (1)

E *Alternative* with more than 15% HGVs entering

Nose (2) and (3)   Entry taper (Slip road right hand lane) (1) Lane width 3.7 m   Ghost island taper (6) Ghost island taper 2 m minimum at widest point

(Layout F is the same as layout E but with both lanes added – the ghost island may be omitted)

N.B. Figures in brackets refer to columns in Table 4/4 of TD 22/92.

Fig. 5.5 Merge with and without lane gain.

Diverge with no lane drop

(A) Taper diverge

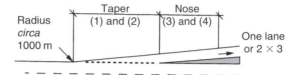

(B) Parallel diverge

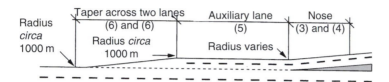

Diverge with lane drop

(C) Mainline lane drop at taper diverge

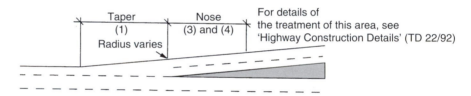

(D) Mainline lane drop at parallel diverge

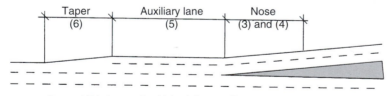

(Layout E is layout D with both lanes off)

N.B. Figures in brackets refer to columns in Table 4/5 of TD 22/92.

Fig. 5.6 Diverge with and without lane drop.

traffic stream separates into two traffic streams. Similarly, the capacity of the diverge is governed by the capacity of the upstream and downstream carriageways or lanes. Merges and diverges are frequently accompanied by priority road markings and signs, such as those on the entries and exits to motorways or expressways. Occasionally, there are more than two entries to a merge or more than two exits at a diverge, but this is unusual and can be hazardous and confusing for drivers.

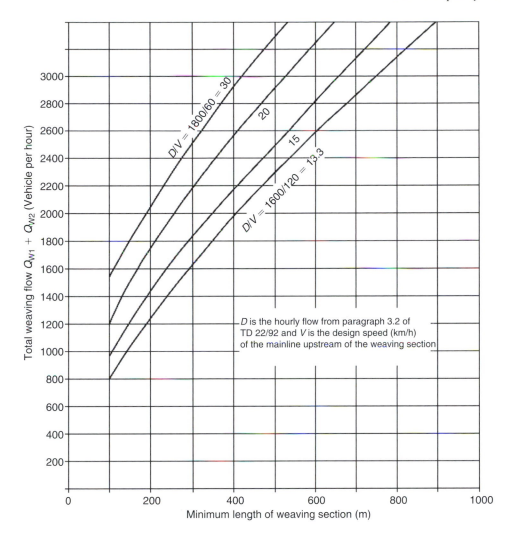

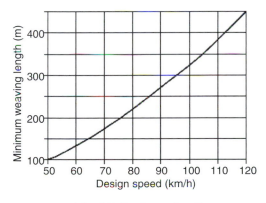

To determine the minimum length of weaving section (L_{min}) for insertion within the formula of paragraph 2.26 of TD 22/92.

1. For known total weaving flow and chosen D/V value, read off the minimum length of weaving section from the graph above.

2. Check the minimum weaving length allowable for chosen design speed from the graph on the left.

3. Select the greater of the two lengths.

Fig. 5.7 Total weaving flow versus minimum length of weaving section.

Crossing conflicts occur where two traffic streams cross each other at an acute angle. Usually, one of the traffic streams has priority over the other. The capacity of the non-priority stream is governed by the speed and intensity of the priority traffic stream. The driver on the non-priority arms of a junction has to observe a suitable gap in the priority stream or streams before entering or crossing the priority stream.

5.8   Merges and diverges

Though merges occur within priority junctions and small roundabouts the 'give-way' control generally requires the minor stream to slow or stop, and wait for a suitable gap, prior to entering the priority stream. On more major roads the merge is provided with an acceleration lane, which enables drivers entering the priority stream to synchronise their speeds and make use of a smaller gap. If the diverge is provided with a suitable length deceleration lane the 'through' priority stream is largely unaffected. Merges and diverges are usually marked with a broken white line to indicate the edge of the priority stream. Figures 5.3 and 5.4 from TD 22/92[9] are used with Figures 5.5 and 5.6, respectively.

5.9   Weaving sections

Weaving sections occur when a merge is followed by a diverge. Weaving sections are common on old style roundabouts and gyratories, and between motorway interchanges. Weaving also occurs in urban streets. The capacity of a weaving section is governed by the length and width of the weaving section, the geometry of the upstream and downstream merges and diverges, and the proportion of weaving and non-weaving traffic in the section (see Figure 5.7 from TD 22/92).

References

1. Transportation Research Board (1985) *Highway Capacity Manual*, Special Report 209, TRB, Washington, DC.
2. Highways Agency (1997) *Assessment and Preparation of Road, Schemes Section 1 Assessment of Road Schemes*, DMRB, Volume 5, Section 1, Part 3, TA 46/97, HMSO, London.
3. Kimber, RM, McDonald, M and Hounsell, NB (1986) *The prediction of saturation flows for road junctions controlled by traffic signals*, TRRL Research Report 67, TRRL, Crowthorne.
4. Webster, FV and Cobbe, FM (1966) *Traffic signals*, Road Research Technical Paper 56, HMSO, London.
5. Office of the Deputy Prime Minister (1996) *Planning Policy Guidance – Town Centres and Retail Developments (PPG 6)*.
6. Office of the Deputy Prime Minister (2001) *Planning Policy Guidance Note 13: Transport (PPG 13)*, March.
7. Department of Transport (1985) *ARCADY2: an enhanced program to model capacities queues and delays at roundabouts*, TRRL Research Report 35, TRRL, Crowthorne.
8. Department of Transport (1985) *PICADY2: an enhanced program to model capacities queues and delays at major/minor priority junctions*, TRRL Research Report 36, TRRL, Crowthorne.
9. Highways Agency (1992) *Layout of Grade Separated Junctions*, DMRB, Volume 6, Section 2, Parts 1 and 2, TA 48/92, HMSO, 1992 and TD 22/92, HMSO, 1992, London.

6

Traffic Signs and Markings

6.1   Introduction

An important part of any road is the means by which the traffic engineering conveys information about the road and any regulations that affects the way it is used to users. If this is done successfully it helps to make travel both safer and more efficient and it helps road users to ensure that they comply with the regulations governing the road that they are using.

In the 21st century technology has reached a point where it is contributing to this process by the use of radio, in-car systems and roadside variable message signs and these are discussed in detail in Chapter 15. However, the primary mechanisms for communicating with the road users remain traffic signs and road markings, discussion of which forms the subject of this chapter.

There are hundreds, if not thousands of different sign designs and it is not possible to list and identify every combination and variation here. In what follows we offer an overview of the current system and some typical examples of the main types of signs. To assist road users the Department for Transport publishes 'The Highway Code'[1] which sets out the main types of signs that the road user could expect to see and it is recommended as a very useful summary and overview of signs and their meaning.

6.2   What is a traffic sign?

Although we colloquially refer to signs and markings, the law does not distinguish between vertical (signs) and horizontal (markings) traffic signs. In the UK the design is very highly regulated with all permissible signs defined precisely in regulations. The current evolution of these regulations is the Traffic Signs Regulations and General Directions 2002 SI 2002/3113.[2] Supplementary guidance as to use and positioning is given in documents such as the Traffic Signs Manual,[2] Numerous Traffic Advisory leaflets and in Circulars such as Circular 7/75 which deals with the size and position of signs.

Traffic signs and markings divide logically into a number of broad types or categories. These are:

- **Warning signs**: As the name suggests provide advanced warning of some feature, such as a low bridge or a road narrowing.
- **Regulatory signs**: Announce and enable traffic regulation. Examples of this type of sign could include a speed limit or a banned movement.

- **Informatory signs**: For example the increasingly common sign which informs motorists of the presence of an enforcement camera ahead.
- **Direction signs**: As the name suggests are the modern day equivalent of the signpost showing the route to be followed to reach a given destination.
- **Road markings**: Cross all groups and can *inter alia* show the position those vehicles should adopt on the road, hazards, the presence of traffic regulations and directional information.
- **Traffic signals**: Are dealt in detail in Chapter 9.
- **Temporary and miscellaneous signs**: The increasing drive for safety, particularly at road works, has generated a complete strategy for signing such works so as to ensure safe traffic movement with signs styles specifically specified for such use. There are also various other one off-sign designs which do not naturally fit into any of the previous categories.

The regulations are very extensive and quite prescriptive on the design of signs. They set:

- Sign size and colour.
- What information is allowed to be displayed, particularly on warning and regulatory signs.
- Illumination requirements.
- The retro-reflectivity of materials to be used in manufacturing the sign face or road marking.

In addition, government guidance sets out factors such as the permitted names and order of precedence for place names on all but local direction signs. All signs must comply with the government's guidelines and if a highway authority wishes to use a non-standard sign they must first get a special approval for the sign from the department who will basically need to be satisfied that (a) a new sign is needed and (b) that the design is harmonised with existing signs.

6.2.1   WARNING SIGNS

Warning signs are typically in the form of a red triangle with the point uppermost and warn of features on the highway, such as low bridge, bend, etc., which the road user needs to be prepared to react to. The centre of the triangle is white with a graphic in black representing the hazard being

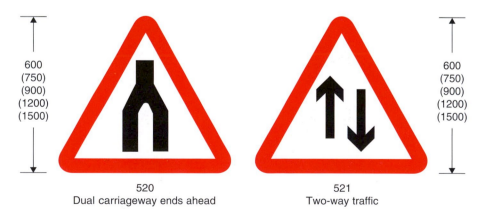

600
(750)
(900)
(1200)
(1500)

520
Dual carriageway ends ahead

600
(750)
(900)
(1200)
(1500)

521
Two-way traffic

Fig. 6.1 Warning signs.

warned about. There are also warning signs such as those showing the likely presence of animals, cyclists, trams, etc., which warn about factors related to the environment and other road users rather than a feature of the highway *per se*. Figure 6.1 shows some typical warning signs.

6.2.2   REGULATORY SIGNS

There is a large variety and mix of sign designs to show regulations. These include speed and weight limits, and banned turns, or prohibited vehicle types. Most of these signs are shown on circular signs with a red border and a white centre showing the regulation or prohibition. Examples of these signs are shown in Figure 6.2. Examples of other regulatory signs include parking regulatory signs shown in Figure 10.2 in Chapter 10.

It cannot be emphasised too strongly the importance of ensuring that these signs are designed and installed *strictly* in accordance with the regulations. Failure to comply with the prescribed design can mean that the signed regulations have no effect. This can mean that a driver who breaks the regulations will escape without penalty and those who assume that they have the protection implied by the 'signed' regulations are actually put at risk.

6.2.3   INFORMATION SIGNS

Information signs give road users information about features and factors which may be of assistance to them in making their journey. These tend to be rectangular blue signs with a white edge and include signs such as the sign showing that a road is a *cul-de-sac*, junction countdown signs and advanced warning signs for features such as a low bridge. In these circumstances the sign would also carry the appropriate triangular warning sign within the blue sign. Some example signs are shown in Figure 6.3.

A new type of information sign has come into use in recent years, that being the variable message dot-matrix signs which are now in quite wide use, particularly on motorways. These signs are described in detail in Chapter 15.

6.2.4   DIRECTION SIGNS

Direction signs are the natural successors of the signpost directing travellers to follow a certain route to reach a given destination. In the UK, we have a hierarchical system of roads and

614
No U-turns for vehicular traffic

615
Priority must be given to vehicles
from the opposite direction

Fig. 6.2 Regulatory signs.

the Direction Signing system broadly reflects this:

- **Motorways** are special roads with access limited to certain types of motor vehicle. They have blue-backed signs.
- **Primary routes** tend to be higher standard roads that connect the main towns and cities, known as 'places of traffic importance'. Primary route signing uses green-backed signs.
- **All other routes** use white-backed signs.

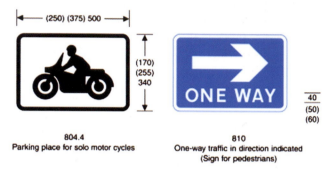

804.4
Parking place for solo motor cycles

810
One-way traffic in direction indicated
(Sign for pedestrians)

Fig. 6.3 Other informatory signs.

Primary route direction sign

Non-primary route direction sign

Pedestrian sign

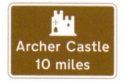

Tourist sign

Fig. 6.4 Direction signing.

- **Tourist signs** are used on all standards of roadway to show the routes to significant tourist attractions, they have a brown background.
- **Pedestrian signing** is generally only used in town centres to show pedestrian routes for local attractions. This signing has a blue background but is unlikely to be confused with motorway signing because of the content and style of the signs.

Examples of all these signs are shown in Figure 6.4.

The biggest challenges facing a local signs engineer are given below:

- Ensuring that signed routes remain coherent and clear. Local highways schemes such as a mini-roundabout can very quickly make it difficult for a stranger to follow; what to a local person is an obvious route. A work in London in the 1980s (ref) showed that, on the capital's primary route network, about two-thirds of the direction signs contained errors.
- Avoiding sign overload, the governments guidance in Circular 7/75 sets out quite clear guidance on how much information a driver might reasonably absorb from a sign, but local needs and local pressures can override these limits. An exercise which sought to simplify and improve signing in Guildford erected new signs which in an attempt to maintain continuity with the approaching routes signed no less than 29 destinations on one approach to a junction!

The government sets out guidance in Circular 7/75 as to the positioning of signs relative to a junction so that the drivers get suitable sight lines and time to read the sign before having to make navigational decisions and manoeuvre. Although this guidance may be followed in rural locations research in London showed that the constraints of an urban road network do not always allow the guidelines to be met.

6.2.5   ROAD MARKINGS

It is not possible to overestimate the importance of road markings as part of the road system. In a few instances, road markings merely emphasise the layout of the highway and guide road users to a safe course of action. In many cases the whole of the success of a scheme relies upon the visual messages emanating from the road markings (e.g. mini-roundabouts are often implemented by road markings alone supplemented by a few traffic signs).

Carriageway markings in the UK must be placed in accordance with the Traffic Sign Regulations and General Directions 2002[2] (TSRGD). Further advice is given in the Traffic Signs Manual.[3]

Road markings must be designed into a scheme at the earliest stage; they cannot be added later when all the other aspects of a scheme have been agreed. Usually the locations of edge of carriageway, lane lines, ghost islands and priority markings are as critical as the location of the kerb lines, traffic islands and other highway features.

Road markings not only guide road users but provide evidence of traffic regulations, such as waiting and loading restrictions, pedestrian crossings, box junctions, keep clear markings and level crossings.

In rural areas the double white-line system uses double longitudinal lines. Solid white lines are used to prevent dangerous crossings of the carriageway centre line. In urban areas solid lines delineate bus and cycle lanes, and stop lines. Yellow lines (and in London and Edinburgh red lines) on the carriageway parallel to the kerb to indicate waiting restrictions; loading restrictions are indicated on the kerb itself. Waiting and loading restrictions markings on kerbs are common in the USA and many other countries.

6.2.6 *TEMPORARY SIGNS*

Increasingly, works on the highway become an extended and have a disruptive effect on traffic flow over a protracted period and this can result in congestion and an increased accident risk. In order to deal with this the government have introduced a system of temporary signing to warn of road works and manage and direct traffic flow. These signs which can include warning regulatory and direction signing have a yellow background upon which the normal signs are superimposed.

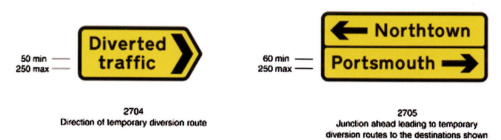

50 min / 250 max

60 min / 250 max

2704
Direction of temporary diversion route

Item	
1	Regulations: None
2	Directions: 13(3)
3	Diagrams: None
4	Permitted variants: Schedule 16, items 15, 20 Any symbol shown in Part VII of Schedule 13 may be substituted for "Diverted traffic"
5	Illumination requirements: Schedule 17, item 4

2705
Junction ahead leading to temporary
diversion routes to the destinations shown

Item	
1	Regulations: None
2	Directions: 13(3)
3	Diagrams: None
4	Permitted variants: Schedule 16, items 3, 5, 12, 16, 19, 20, 30, 31, 32, 34
5	Illumination requirements: Schedule 17, item 4

60 min / 250 max

50 min / 250 max

2706
Roundabout ahead leading to a temporary
diversion route to the destination shown

Item	
1	Regulations: None
2	Directions: 13(3)
3	Diagrams: None
4	Permitted variants: Schedule 16, items 3, 5, 16, 19, 20, 30, 31, 32, 34
5	Illumination requirements: Schedule 17, item 4

2707
Direction of temporary diversion route
to destination shown

Item	
1	Regulations: None
2	Directions: 13(3)
3	Diagrams: None
4	Permitted variants: Schedule 16, items 3, 5, 15, 19, 20, 30, 31, 32, 34
5	Illumination requirements: Schedule 17, item 4

Fig. 6.5 Temporary signs.

They can also be used to show permanent changes in traffic flow pending the erection of permanent signing. Generally speaking the standards of reflectivity and lighting are lower than for permanent signing. Examples are shown in Figure 6.5.

References

1. "The Highway Code" the Department for Transport, London.
2. UK Government (2002) The Traffic Signs Regulations and General Directions SI 2002/3113, HMSO, London.
3. Department for Transport (1974) The Traffic Signs Manual (separate chapters 1–14 not all available), HMSO, London.

7

Traffic Management and Control

7.1   Objectives

Traffic management arose from the need to maximise the capacity of existing highway networks within finite budgets and, therefore, with a minimum of new construction. Methods, which were often seen as a 'quick fix', required innovative solutions and new technical developments. Many of the techniques devised affected traditional highway engineering and launched imaginative and cost-effective junction designs. Introduction of signal-controlled pedestrian crossings not only improved the safety of pedestrians on busy roads but improved the traffic capacity of roads by not allowing pedestrians to dominate the crossing point.

More recently the emphasis has moved away from simple capacity improvements to accident reduction, demand restraint, public transport priority, environmental improvement and restoring the ability to move around safely and freely on foot and by pedal cycle.

7.2   Demand management

There has been a significant shift in attitudes away from supporting unrestricted growth in highway capacity. The potential destruction of towns and cities and the environmental damage to rural areas is not acceptable to a large proportion of the population. Traffic management has, largely, maximised the capacity of the highway network yet demand and congestion continues to increase.

Highway authorities accept that they do not have a mandate to provide funds for large amounts of new construction. It is clear that, for the foreseeable future, resources will not be available to provide for unrestricted growth in private vehicular traffic. Traffic engineering alone cannot provide sufficient highway capacity even with limited amounts of new construction.

As traffic demand and congestion increased, drivers found alternative routes, often though residential areas. Road safety was compromised as drivers travelled at high speed to maximise the benefits of diverting from their normal route. Pressure from residents, in these areas, led to the introduction of area-wide Environmental Traffic Management Schemes (ETMS) during the 1970s and 1980s.

ETMSs attempted to deny these 'rat-runs' to queue-jumping traffic and to specific classes of vehicle such as wide or heavy vehicles. Many ETMSs were spectacularly effective and used such techniques as point road closures, physical width restrictions, one-way 'plugs', one-way streets and banned turns. These measures, which were designed to be restrictive for unwelcome traffic, often caused great inconvenience to residents, emergency vehicles and service vehicles. Frequently residents were prepared to tolerate severe inconvenience in order that a safe and tranquil environment could be restored.

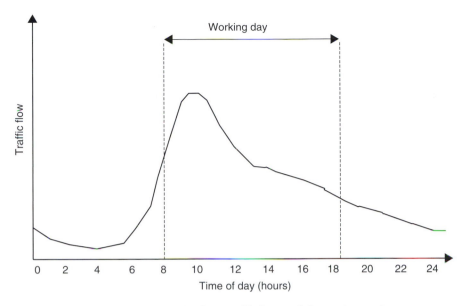

Fig. 7.1 Typical daily flow profile for a radial route into a city.

In recent years the emphasis has moved towards reducing vehicle speeds and accidents using traffic calming and reallocation of road space. ETMS techniques are still widely used often in combination with traffic calming methods. Often the broad aims of ETMSs can be achieved by traffic calming alone. Traffic calming is discussed in detail in Chapter 12.

In town centres whole carriageways have been closed to general traffic and dedicated to pedestrians, cars driven by people with disabilities, buses and service vehicles. Many town centre schemes were preceded by hostile objections from shopkeepers concerned about loss of passing trade. However, most town centre schemes have produced a much improved trading environment. Outside shopping hours many centres appear lifeless and, in some instances, general traffic has been allowed back into the centre during those hours.

Figure 7.1 shows a typical daily flow profile for a radial route into a large town or city. It can be seen that peak periods are relatively short and that for long periods of the day traffic flows are below the road's capacity.

As delays increase, drivers realise that there is spare capacity at other times of the day and change their working hours to arrive at and depart from their places of work before and after the peaks. The effect of this is that peaks last longer and is known as 'peak spreading'. Ultimately the peaks spread until the morning and afternoon peaks meet. The daily flow profile in Figure 7.2 shows the majority of daytime capacity has been utilised. This effect is observable in parts of Central London and other large cities where heavy congestion occurs throughout the working day. Sunday trading has produced dramatic increases in traffic flows on what was a lightly trafficked day. The only remaining part of the day where capacity is available is during the night; even this is now under attack as some large superstores are staying open throughout the 24-hour day.

In many cities overall journey speeds are restricted to around 10 mph and have a severe effect upon the economic health of the centre, the physical health of the population and essential traffic such as the emergency services and public transport are severely impeded.

It is clear that unrestricted use of private cars cannot continue. Some limitation on the freedom of use is inevitable if towns and cities are not to become polluted wastelands clogged with slow moving traffic for most of the day.

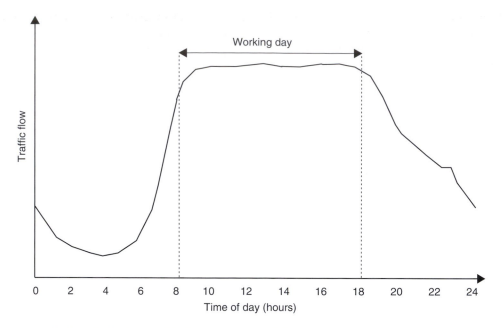

Fig. 7.2 Daily flow profile showing capacity fully utilised during the working day.

Removal of the need to travel by private car and removal of demand by means of parking and junction control with parallel incentives to use public transport offer a solution.

One of the most effective ways to restrict private car traffic to a city centre is on- and off-street parking control. Parking control is discussed in Chapter 9. Restriction on numbers together with punitive charges forces drivers to consider alternative modes for their journeys. Control of private non-residential (PNR) parking must accompany control of public car parks and on-street parking or many of the benefits will be lost.

The town planning and development control process must keep pace with parking policy, and there might be a need for additional legislation to ensure that neighbouring authorities produce compatible policies.

Reduction in highway capacity by transferral of road space to more efficient use by the high capacity vehicles such as buses and trams is doubly effective. Bus lanes reduce journey times and provide a clear demonstration to car drivers of the efficiency of public transport. If a significant modal shift occurs bus journey times reduce further.

Reduction in highway capacity must be applied over a whole network or untreated, unprotected routes will be used by drivers as alternative routes. This effect is similar to the rat-running problems encountered in the early days of ETMS. The wider area effects of such measures can be tested using traffic models such as TRIPS and SATURN and the micro-simulation models Paramics, VISSIM and AIMSUN. Sometimes, where the network being considered is relatively compact, carefully monitored experiments are more productive. In large conurbations and major cities diversionary effects can occur at a great distance from the site under consideration. Drivers discouraged from using a radial route may divert to another radial, using a sub-orbital route (e.g. M25 in London). In most cases the 'settling down' or stabilising of traffic flows can take many weeks. Local congestion caused by an experimental restraint measure can create intense local pressure for its removal and before the new traffic patterns have been established.

One method of demand management that has received considerable interest and study is congestion charging. This is where vehicles are surcharged for their use of road space depending on the prevailing levels congestion. New technology in the form of smart cards and vehicle identification are needed to ensure that the system is practical and fair. Measures to accommodate 'foreign' or non-local vehicles are also needed. Video image analysis has reached a level of sophistication that can be used for this purpose.

Detailed congestion charging experiments and studies have been carried out in London[1] and Cambridge. In February 2003 the Mayor of London introduced congestion charging in central London. Drivers are licensed to drive anywhere within London's inner ring road for a charge of £5. The scheme has been deemed a success as general traffic flows have been reduced by approximately 15% as predicted by the London Transportation Study multi-modal model. Private car traffic has been reduced by more than 35% in the peak periods. The introduction of charging was accompanied by an increase in bus service provision and ridership. All people, particularly those who work unsocial hours and motoring groups, do not universally applaud congestion charging. However, the Mayor has initiated studies into possible extensions of the charging area. Other major cities in the UK are actively studying congestion charging or area licensing, the most notable being Edinburgh.

7.3   Engineering measures

The traffic engineer can apply an extensive array of measures to achieve his objectives. These objectives include:

- capacity enhancements;
- accident reduction;
- environmental protection and enhancement;
- servicing of premises and providing access;
- providing assistance to pedestrians and cyclists;
- assisting bus or tram operators;
- providing facilities for persons with disabilities;
- regulating on- and off-street parking.

The majority of capacity problems occur at road junctions. In urban areas road junctions are important focal points for pedestrian and cycling activity and are often the site of public transport interchanges. Due to the various conflicting demands it is not surprising that two thirds of urban traffic accidents occur at road junctions. Selection of an appropriate junction design for a particular site can be very difficult. Some designs, such as roundabouts, can significantly reduce the severity of vehicle–vehicle accidents but can prove hazardous for cyclists. In some instances installation of traffic signals with full pedestrian and cycle facilities and bus priority measures might also reduce the overall traffic-handling capacity.

Careful allocation of road space to separate traffic streams into designated traffic lanes can reduce confusion and limit accidents. Designation of traffic lanes might include special vehicle lanes, such as cycle and bus lanes and dedicated left or right-turn lanes.

Introduction of banned turns and one-way streets can reduce potential conflicts and accident potential. These measures can be used to implement protected pedestrian or cycle crossings and simplify junction layouts generally. Great care must be exercised when one-way street schemes are being considered, as they can result in speeding by drivers who are confident that they will not be opposed by other vehicles.

Point road closures are used to simplify junction and highway layouts and eliminate turning conflicts. The resulting continuous footway can also improve pedestrian safety and provide space for bus stops, cycle racks, pedestrian crossings and hard and soft landscaping.

Closure of long sections of road to general traffic can produce pedestrianised shopping streets. Such schemes can be very complex, to design and introduce, as facilities for buses, emergency services, residents/proprietors and service vehicles must be considered.

In residential streets, a more domestic scale to the local environment can be achieved by the introduction of Homezones. In these schemes residents participate in the re-design of their streets to reduce traffic speed, discourage through traffic, improve safety and the appearance of the area. Homezones are development of methods used elsewhere in Europe, most notably the Woonerf schemes in Holland.

Carriageway narrowing can be used to limit capacity or vehicle speeds and reduce parking and pedestrian crossing distances. Carriageway narrowing is discussed in Chapter 12.

The key to all successful traffic engineering schemes is that the visual cues provided by the road must give a clear indication to users of who has priority.

7.4   Junction types

There are many varying junction types, in detail, but they can be broken down into five basic types.

1. Uncontrolled non-priority junctions
2. Priority junctions
3. Roundabouts
4. Traffic signals
5. Grade separation.

Each junction type can accommodate different levels of major and minor road flow. Choice of a particular junction type depends upon the flow levels and the space available for its construction or reconstruction (see Figures 7.3–7.7).

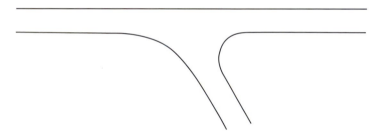

Fig. 7.3 Uncontrolled non-priority junction.

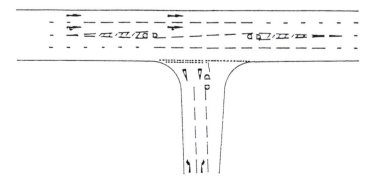

Fig. 7.4 Priority junction.

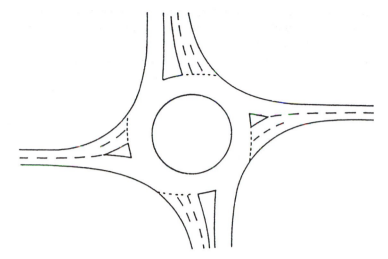

Fig. 7.5 Roundabout.

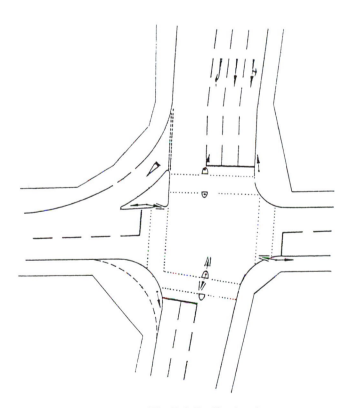

Fig. 7.6 Traffic signals.

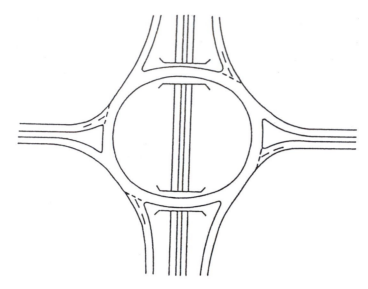

Fig. 7.7 Grade separation.

7.4.1   UNCONTROLLED NON-PRIORITY JUNCTIONS

At the lowest flow levels, such as those experienced in small residential areas, non-priority junctions are suitable. However, even at very low flow levels, a lack of priority can lead to confusion and accidents. For modest costs, simple priority road markings can remove uncertainty.

7.4.2   PRIORITY JUNCTIONS

These vary from very simple 'T' junctions, with 'give-way' markings, catering for low flows to highly complex junctions on single or dual-carriageway roads, with turning movements separated by channelisation islands, 'ghost islands' and auxiliary priority markings and signs within the overall junction area. If sufficient space is available, very heavy vehicular flows and high speeds can be handled safely.

7.4.3   ROUNDABOUTS

Roundabouts can be considered as a special type of priority junction. Vehicles 'give-way' to offside traffic and circulate around a central island (in a clockwise direction in the UK). Roundabouts can be very large, up to 200 metres or more across and as small as 13 metres across.

The smallest roundabouts are called mini-roundabouts and the central island is reduced to a raised, drive-over dome or a simple flat circle, usually painted white and accompanied by a series of 'circulating' arrows marked around the centre.

Though the very small and large roundabouts both have offside priority markings they function quite differently. The circulating movement is almost absent in the smaller junctions and, 'straight' across movements become crossing conflicts. Roundabouts were a particularly British device but are now in common use in most countries; France in particular has introduced them widely now that they have resolved their nearside priority problems (i.e. traffic entering has to give way to traffic already on the roundabout).

7.4.4   TRAFFIC SIGNALS

Traffic signals were originally used when there was insufficient land available for enlarging a junction. Increasing sophistication in the control systems has resulted in their widespread use in heavily trafficked urban areas. Positive control is now applied to different road users such as pedestrians, cyclists and buses.

Traffic signals can operate on fixed-time plans where the time allowed for each arm is preset, using historic traffic data. Vehicle actuation uses detectors to measure traffic demand on the approaches to the junction and vary the amount of green time allotted to each arm. Green time can be varied between a minimum and maximum time that is preset in the controller. Under conditions of low flow, detection of a vehicle can initiate a green stage. The most sophisticated form of local adaptive control is the MOVA system developed by the UK Transport Research Laboratory.

There are many examples where signals have been installed on roundabouts with severe congestion problems. For many years traffic signals have been linked over wide areas to minimise delay and maximise network capacity. Linking can be simple synchronisation of successive sets of signals to produce a green wave for vehicles along a route, fixed-time plan Urban Traffic Control and adaptive systems such as SCOOT[2] have been used to maximise network capacity for many years.

7.4.5   GRADE SEPARATION

Where a particular movement through a junction has a very high flow it can be separated vertically from the other turning movements. The simplest grade separation, at a cross roads, is the diamond interchange. The minor road passes under or over the major road and slip roads connect to the major road from an at-grade junction on the minor road. Depending on the flow levels, the minor at-grade junction could be controlled by a priority junction, a roundabout or roundabouts or traffic signals.

Where two or more major highways cross, the connections are often made by connecting the slip roads together and avoiding at-grade conflicts altogether.

Grade separation is used on all motorways in the UK and on other major highways. Some examples of grade-separated junctions are shown in Figure 7.8.

In recent years the UK Highways Agency has piloted the use ramp metering on motorways. Ramp metering uses traffic signals to control the entry of traffic from motorway slip roads onto the main carriageway and, thus, minimise flow breakdown. With increasing travel demand ramp metering is one of a number of measures aimed at optimising the safe use of the existing highway network. Ramp metering is used successfully in France, Holland and the USA.

7.5   Road markings

It is not possible to overestimate the importance of road markings as part of the road system. In a few instances road markings merely emphasise the layout of the highway and guide road users to a safe course of action. In many cases the whole of the success of a scheme relies upon the visual messages emanating from the road markings, (e.g. mini-roundabouts are often implemented by road markings alone supplemented by a few traffic signs).

Road markings must be designed into a scheme at the earliest stage; they cannot be added later when all the other aspects of a scheme have been agreed. Usually the locations of edge of carriageway, lane lines, ghost islands and priority markings are as critical as the location of the kerb lines, traffic islands and other highway features.

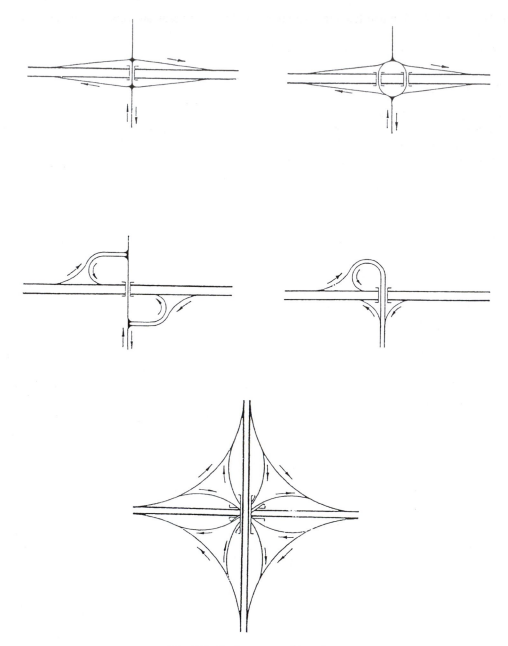

Fig. 7.8 Grade-separated junctions.

Road markings not only guide road users but provide evidence of traffic regulations such as waiting and loading restrictions, pedestrian crossings, box junctions, keep clear markings, level crossings.

In rural areas the double white line system uses double longitudinal lines. Solid white lines are used to prevent dangerous crossings of the carriageway centre line. In urban areas solid lines delineate bus and cycle lanes and stop lines. Yellow lines (and in London red lines) on the carriageway parallel to the kerb to indicate waiting restrictions; loading restrictions are indicated on the kerb itself. Waiting and loading restrictions markings on kerbs are common in the USA and many other countries.

Road markings are discussed in more detail in Chapter 6.

7.6   Traffic signs

Traffic signs also provide a vital function for road users, warning signs provide information to road users about hazards such as: junctions, changes of direction, carriageway width, gradient, low, opening or humped back bridges, roadworks, etc.

Regulatory signs provide a message that must be obeyed, for example stop, 'give-way', banned turns, compulsory turns, no entry, one-way streets, prohibited vehicle types, weight and width restrictions, waiting and loading restrictions, speed restriction, etc.

Directional informatory signs provide information about routeing, important places of interest such as railway stations, airports, etc.

Other informatory signs provide information about footway and other parking schemes, heritage sites, census points, etc. Traffic signs are often installed in association with road markings (e.g. 'give-way' triangle and waiting and loading plates).

In recent years variable message signs (VMS) have become more common. These signs are controlled by car park entry systems, motorway control centres, Urban Traffic Control centres and, more recently, by speed cameras, CCTV and video imaging systems. There are two basic types of VMS in common use in the UK: rotating boards and dot-matrix signs. Rotating board signs can have double sided or triangular prism signs. These signs are clear and easy to read as they appear very similar to static road signs; their main disadvantage is that they are restricted to two or three fixed messages. Dot-matrix signs use fibre optics to display a wide variety of messages and are, thus very flexible in operation. However the operator must be careful to use easily read unambiguous messages. Modern signs are reasonably easy to read except in conditions of direct, bright sunlight. Recently there have been moves to use sponsored dot-matrix signs for commercial advertising when they are not needed for traffic purposes. This use might have road safety implications.

Traffic signs are discussed in more detail in Chapter 6.

References

1. Gilliam, C and Richards, M (1996) The MVA Consultancy 'The London Congestion Charging Research Programme', a series of papers published in *Traffic Engineering & Control* **37**(2), 66–71; **37**(3),178–179, 181–183; **37**(4), 277–282; **37**(5), 334–339; **37**(6), 403–409.
2. Department of Transport (1981) '*SCOOT: A Traffic Responsive Method of Coordinating Signals*', TRRL Laboratory Report 1014, TRRL, Crowthorne.
3. UK Government (2002) *Traffic Signs Regulations and General Directions 2002*, HMSO, London.
4. UK Government (1974) *The Traffic Signs Manual* (separate Chapters 1 to 14 not all available), HMSO, London.

8

Highway Layout and Intersection Design

8.1 Highway link design standards

It is unlikely that the traffic engineer alone will be required to design major highways. Highway design is a separate, albeit a closely related, discipline. At most the traffic engineer will provide preliminary layouts of access roads and intersections for developments and, therefore, a good understanding of basic highway design techniques and standards is needed.

8.1.1 DESIGN SPEED

Selection of an appropriate design speed may be considered the starting point for any scheme. The DMRB Technical Standard TD 9/93[1] outlines the methods for selecting the link design speed.

In practice, the design speed for a particular route might be contained within a policy decision by the highway authority (HA). The engineer should consult the HA closely and, if no advice is forthcoming, suggest a method for determining a suitable design speed. In which case, the design speed should be selected by observation of the actual behaviour of the vehicles on the road in question.

Vehicle speeds are affected by many factors including speed limit, horizontal and vertical alignment, visibility, highway cross section, adjacent land use, spacing of junctions, accesses, pedestrian crossings and maintenance standards. The general condition and design of vehicles and driver ability, which changes over time can have a significant effect on vehicle speeds. It is usual to use an 85th percentile speed as the design speed (the speed below which 85% of drivers travel). The 85th percentile speed is determined from speed surveys using a radar speed meter or automatic traffic counters equipped to measure speeds.

In urban areas, where the speed limit is less than the national speed limit, the design speed is more likely to be based upon a policy decision. Frequently the traffic engineer is asked to reduce the speeds of vehicles to an acceptable level. In these circumstances it would be wholly inappropriate to use the 85th percentile speed as the design speed as this is often well above the existing or proposed speed limit. The engineer may be instructed to introduce measures that reduce or control 85%, or an even higher percentage, of vehicle speeds to the speed limit.

8.1.2 HORIZONTAL CURVATURE

The radius of curvature that a vehicle can travel round at the design speed depends upon the crossfall or cant of the carriageway and adhesion between the tyres and the road surface. The highways agency provides a table of standards, Table 8.1,[1] for standard design speeds for different crossfalls or superelevation. Figure 8.1 shows the forces acting upon a vehicle travelling around a curve.

Table 8.1 Recommended Geometric Design Standards

Speed limit		Design speed
MPH	KPH	KPH
30	48	60B
40	64	70A
50	80	85A
60	96	100A

Source: DMRB TD 9/93.

Design speed (kph)	120	100	85	70	60	50	V^2/R
Stopping sight distance (m)							
Desirable minimum	295	215	160	120	90	70	
One step below desirable minimum	215	160	120	90	70	50	
Horizontal curvature (m)							
Minimum R* without elimination of	2880	2040	1440	1020	720	520	5
Adverse Camber and Transitions							
Minimum R* with superelevation of 2.5%	2040	1440	1020	720	510	360	7.07
Minimum R* with superelevation of 3.5%	1440	1020	720	510	360	255	10
Desirable minimum R* with superelevation of 5%	1020	720	510	360	255	180	14.14
One step below desirable minimum R with superelevation of 7%	720	510	360	255	180	127	20
Two steps below desirable minimum radius with superelevation of 7%	510	360	255	180	127	90	28.28
Vertical curvature							
Desirable minimum* crest K value	182	100	55	30	17	10	
One step below desirable min crest K value	100	55	30	17	10	6.5	
Absolute minimum sag K value	37	26	20	20	13	9	
Overtaking sight distances							
FOSD (m)	*	580	490	410	345	290	
FOSD overtaking crest K value	*	400	285	200	142	100	

* Not recommended for use in the design of single carriageways.
FOSD: Full overtaking sight distance.
'K value': used to calculate the lengths of vertical curves in paragraph 7.1.12.

The centrifugal force acting on the vehicle

$$\frac{Mv^2}{R}$$

where:

   M = mass of the vehicle
   v = velocity in metres/second
   R = radius of curvature

The force in the vertical direction
   Mg (g = acceleration due to gravity.)

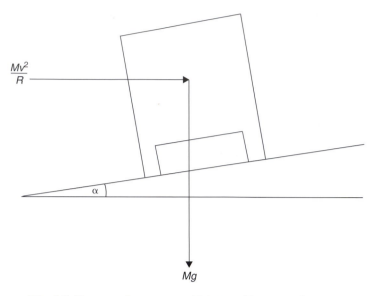

Fig. 8.1 Forces acting upon a vehicle travelling around a curve.

If the forces acting parallel to the road surface are to equal each other then:

$$\frac{Mv^2}{R} \cos \alpha = Mg \sin \alpha$$

therefore:

$$\tan \alpha = \frac{v^2}{gR} \qquad \alpha = \text{the superelevation (cant or crossfall).}$$

In practice these forces are rarely in perfect balance and the differences are accommodated in the friction between the tyres and the road surface.

Generally the desirable minimum radius should be used for new construction. If there are valid cost or environmental reasons, a relaxation of standards to the absolute minimum radius can be permitted. The limiting radius can only be used under certain circumstances at difficult sites. Departures from standards are not recommended except in urban areas where full standards cannot be achieved. In these circumstances it might be possible to use traffic engineering and traffic calming techniques to mitigate the effects of sub-standard alignment (Chapter 12).

8.1.3   TRANSITION CURVES

A transition curve varies in radius from the straight line to the horizontal curvature of the road. Various curve forms have been used such as a true spiral, the cubic parabola and the lemniscate. The rate of increase of centripetal or radial acceleration should normally be limited to 0.3 metres/second³ but can be increased to 0.6 metres/second³ at constricted sites. Where superelevation is to be introduced or adverse camber eliminated, it should be done over the length of the transition curve. The transition curve starts approximately half its length before the normal circular curve tangent point and 'shifts' the circular curve inwards towards its origin (Figure 8.2).

Superelevation improves safety and comfort within the vehicle and allows use of smaller radii. In urban areas superelevation should not, and often cannot, be applied slavishly. Road levels are usually

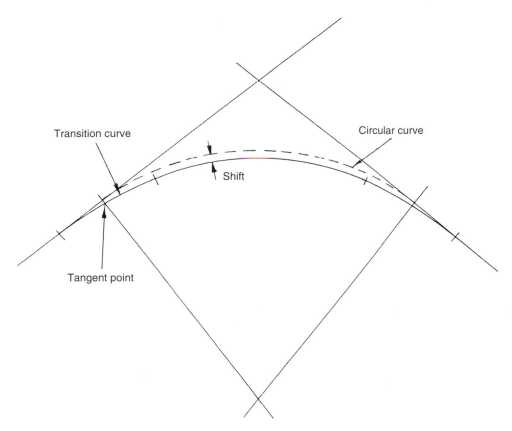

Fig. 8.2 Transition curve.

constrained by frontage development and road junctions. In urban areas it is often more appropriate to provide a camber across the cross section of the road that is sufficient for drainage purposes only.

Within road junctions superelevation should be avoided where possible and should **always** be avoided at roundabouts where it encourages excessive speeds and can seriously affect visibility and understanding of the layout by drivers (Section 8.3).

8.1.4   VERTICAL ALIGNMENT

Vertical alignment consists of a series of straight gradients and connecting curves. The main constraints are comfort and visibility over crest curves. Climbing vehicles are slowed by steep gradients, which should be avoided wherever possible. If long steep gradients cannot be avoided consideration should be given to the provision of crawler lanes for heavy vehicles. In urban areas, gradient is often constrained by side roads and buildings. In many towns and cities very flat roads may force the introduction of artificial gradients or 'false channels' simply to provide adequate surface run off.

The length of vertical crest and sag curves are calculated using the 'K' value obtained from Table 8.1[1] and the difference in the percentage gradients. NB: A rising gradient from left to right is positive and a falling gradient is negative.

Length of vertical curve $L(\text{m}) = KA$

where: K = design speed coefficient
A = difference in percentage gradients.

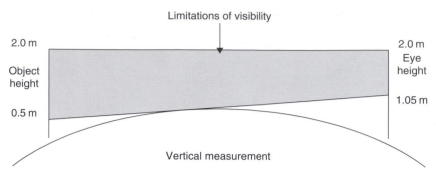

Fig. 8.3 Measurement of SSD.

8.1.5   SIGHT DISTANCES

Table 8.1 gives two sight distances:

- Stopping sight distance (SSD).
- Full overtaking sight distance (FOSD).

SSD is the distance required by drivers to stop their vehicles if they see an unexpected obstruction on the road surface. The SSD includes the 2 seconds driver reaction time and the braking distance on a wet road.

The vertical SSD envelope is shown on Figure 8.3. The driver's eye height is between 1.05 and 2.0 metres above the road surface.

The horizontal SSD envelope is measured from the centre of the nearside lane across the bend. To obtain the envelope the SSD is measured at points around the curve from positions where the sight line lies within the carriageway. On very tight curves, the need to acquire additional land (land take) for SSD can be considerable and sometimes the provision of a larger radius of curvature results in less land take.

On rural single carriageways it is sometimes necessary to provide FOSD. The details of its application are given in TD 9/93.[1] It is unlikely that the majority of traffic engineers will need to design a scheme with FOSD, especially in urban areas.

In urban areas there is immense pressure to place street furniture on footways and verges that could compromise the SSD.

8.1.6   INTERSECTIONS: GENERAL

In rural areas there is often space to provide generous layouts for new junctions. Land cost is relatively low and it is usually possible to adjust the overall scheme layout to achieve full design standards. There are, of course, constraints such as the difficult terrain found in mountainous or hilly areas that physically restrict the freedom of the engineer to optimise designs. There are other constraints that have increased in importance in recent years such as:

- areas of outstanding natural beauty (AONB);
- sites of special scientific interest (SSSI);
- listed buildings and conservation areas;
- archaeological sites;
- high quality farm land.

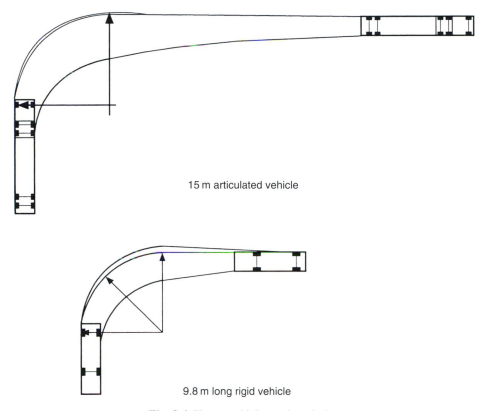

15 m articulated vehicle

9.8 m long rigid vehicle

Fig. 8.4 Heavy vehicle turning circles.

Even with these constraints, it is important and mostly possible to achieve close adherence to design standards. Design speeds in rural areas are usually higher than in urban areas, and the consequences of accidents and collisions at higher speeds are more severe.

With urban areas the physical constraints are more related to the built environment. Property values are higher and opportunities to appropriate land for highway improvements are fewer. Even within the highway boundary the presence of underground services can impose conditions upon the layout. With some small traffic schemes, the cost of protecting or re-routeing services can be many times more than the cost of the highway and signing works.

With any urban scheme the most important consideration is the size, turning characteristics and space requirements of the design vehicles. In most instances this means the largest legally permitted road vehicles. In residential areas the predominate vehicle will be the private car, but large vehicles regularly penetrate these areas for refuse collection, gulley cleaning, maintenance, household deliveries, school buses and public transport. In town centres, department stores, supermarkets, public houses and even fast food outlets receive regular deliveries by the largest heavy goods vehicles. Figure 8.4 shows typical large vehicle turning circles.[3]

Occasionally abnormal loads enter town and city centres. Major cities, such as London, have designated abnormal load routes that avoid weak or low bridges, difficult turning manoeuvres, or overhead telephone or power lines. The engineer must be aware of these and ensure that provision is made, within new designs, for these eventualities. The additional cost of installing hardened over-run areas and de-mountable street furniture can sometimes be recovered from the operators of these outsized vehicles.

In very rare cases the design vehicle can be limited to less than the largest vehicles where physical width or weight limits exist. Where general traffic flows and HGV movements are low, the occasional large vehicle can be allowed to dominate the full width of a road for the short time it needs to complete its manoeuvres.

At urban road junctions it is often not possible to provide full standards for large vehicles. Where their numbers are small and speeds are low they can be allowed to dominate two or more traffic lanes at urban roundabouts and traffic signals. Traffic islands, kerb lines and street furniture must, of course, be placed outside the vehicle swept paths. Standard vehicle swept path plots are available for checking layouts but for more complex multiple turns the DoT computer program TRACK[2] (also AutoTrack) is recommended.

Apart from abnormal load vehicles the longest vehicle in every day use in the UK is the 18-metre drawbar trailer combination and the 18-metre articulated bus. When designing junctions at confined sites, it is not enough to check that the layout is adequate for the largest vehicles. Swept path is dependent on a number of factors:

- rigid or articulated;
- wheelbase (or tractor wheelbase);
- front and rear overhang;
- width;
- length of trailer.

In some instances an articulated vehicle has a smaller turning circle and smaller swept area than a long rigid vehicle. Vehicles with long front or rear overhangs such as the 12-metre European Low Floor Bus and the 18-metre articulated bus can pose particular problems. If such a vehicle is stationary and its steering wheels are on full lock the rear will tend to move outwards as it starts to move. If the vehicle is close to the kerb the bodywork might strike a pedestrian.

8.1.7   SELECTION OF JUNCTION TYPE

Selection of junction type can be very simple and obvious in some cases, for example:

- Two lightly trafficked residential roads – priority junction;
- The through carriageway of a motorway – grade separation;
- Heavily trafficked urban cross roads with heavy pedestrian flows – traffic signals;
- Suburban dual carriageways with substantial heavy goods traffic – conventional roundabout.

However, there are very many cases where the solution is anything but obvious and the engineer should resist making a decision until all the evidence has been gathered, examined and analysed.

Frequently, when an existing junction is to be upgraded to handle more or different types of vehicle, the existing form or control method must be considered. The form of the other junctions on the main route might determine the form of the new or improved junction, (e.g. it might be inappropriate to construct a mini-roundabout on a route that has a series of linked signalled junctions and pelican crossings). Similarly a signalled junction on an otherwise free-flowing rural dual carriageway with generously designed priority junctions or roundabouts could prove to be hazardous.

All junction design is an iterative process that moves between space and capacity requirements. A good starting point is to look at all the required turning movements and their design flows. A conflict diagram as shown in Figure 8.5 is invaluable. The traffic flows can be superimposed on the conflict diagram to give an indication of the importance of each conflict (Figure 8.6). If possible the junction movements should be reduced to a series of three arm 'T' junctions. Some

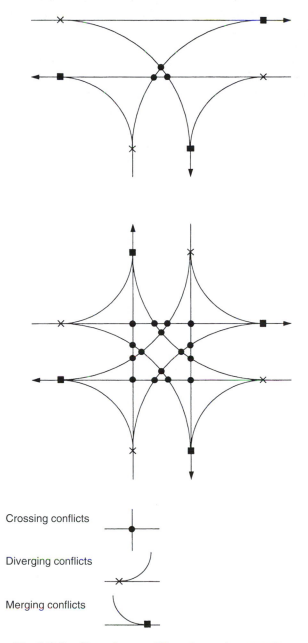

Fig. 8.5 Conflict points at a T junction and crossroads.

minor movements can be prohibited if there are suitable alternative routes for the prohibited traffic, (e.g. cross roads becomes right/left staggered 'T' junctions). Conflict diagrams should be sketched for all extremes of traffic flow, for example, morning and evening peak periods or perhaps heavy weekend flows into a retail park or a sporting event. As these flows might have completely different turning proportions, it is not sufficient to design for the highest throughput at the junction.

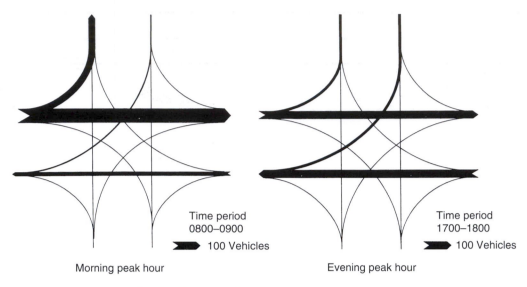

Fig. 8.6 Typical peak period junction flow diagrams.

8.2   Priority junctions

Priority junctions are the simplest and most common of intersections and range from single lane approach 'T' junctions to high capacity channelised layouts. The control at priority junctions depends upon 'give-way' road markings and post-mounted signs.

Cross-road layouts should be avoided wherever possible as they concentrate a large number of vehicle movements and, therefore, conflicts within the junction. On unlit rural roads at night, drivers on the side roads can be confused into thinking that they are on the major route, particularly when there is an approaching vehicle on the far side of the cross roads.

The DMRB Volume 6 Section 2 Part 6 TD 42/95[3] provides details on the layout of major/minor junctions. The TD 42/95 layouts will accommodate the largest vehicle swept paths and a simple check by the engineer will confirm that a design is satisfactory. In urban areas it might not be possible or desirable to achieve the full standards and more detailed checking of the layout will be necessary. If the generous turning radii are provided in areas of extreme parking pressure, motorists might be tempted to park within the junction area. This will compromise the layout and interfere with the swept paths. Waiting and loading restrictions or kerb buildouts can be used to limit this problem.

8.2.1   VISIBILITY

To enable drivers to safely cross or enter the major road traffic streams, they must be able to see and to judge their approach speeds and available gaps in the major road traffic.

Visibility splays are provided from the side road to left and right. A triangular sight line envelope, measured along the centreline of the side road and along the major road kerbline must be maintained clear of obstruction above the drivers eye-height of 1.05 metres. The visibility envelope is defined by its 'x' and 'y' distances (Figure 8.7).[4] The 'x' distances should be a maximum of 15.0 metres and a minimum of 9.0 metres at junctions on major roads. The 'x' distance can be reduced to 4.5 metres where flows are low and an absolute minimum of 2.4 metres. The 'y' distance is dependent upon the major road design speed and its distance should not be compromised. The

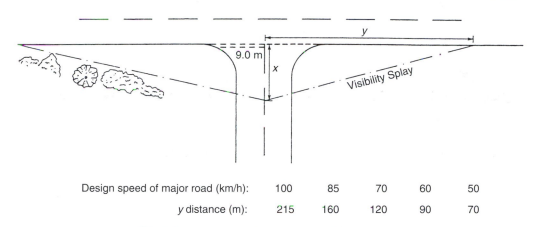

Design speed of major road (km/h):	100	85	70	60	50
y distance (m):	215	160	120	90	70

Fig. 8.7 Visibility requirements at a priority junction.

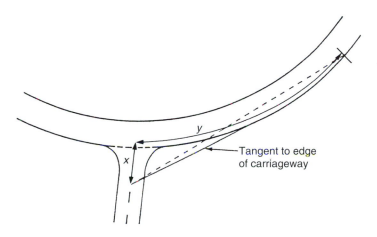

Fig. 8.8 Visibility requirements at a priority junction on a curved major road.

visibility splay joins the two points defined by '*x*' and '*y*' except on a curved alignment where it should be tangential to the carriageway edge (Figure 8.8).[4] Generally the visibility splay should be within the highway boundary to ensure that it can be maintained free of obstruction. Highway authorities can exercise powers under the Highways Act to maintain clear sight lines within the curtilage of private property where necessary.

Where the major road has been provided with a central reservation, or 'ghost island' for vehicles to wait, the visibility distance should also be provided.

At larger, more heavily trafficked junctions or where there are significant pedestrian movements, guide islands and pedestrian refuges can be provided. Channelisation islands help to separate movements and reduce the number of simultaneous decisions required of drivers and pedestrians. For very heavy pedestrian movements formal pedestrian crossings are sometimes installed. The crossings should be located at a suitable distance from the main junction to ensure that drivers can see pedestrians clearly and queuing vehicles from the crossing do not seriously impede movement through the junction.

Pedestrian routes should avoid the most hazardous areas within the junction and should lead to suitably safe crossing points, with short crossing points that are not obstructed by parked vehicles. Pedestrian safety barriers (guard railing) and footway bollards can be used to prevent dangerous crossings and to guide pedestrians to the safety crossing points.

8.2.2   PRIORITY JUNCTION CAPACITY ASSESSMENT

The TRL Computer program PICADY[4] (*Priority Intersection Capacity And DelaY*) uses empirical formulae to model capacities and time-dependent queuing theory to estimate queues and delays for three and four arm major/minor junctions and left–right and right–left staggered cross roads. The program requires the input of demand traffic and pedestrian flows (where pedestrian crossings are present) and various geometric parameters which describe the junction. The program can simulate peak traffic conditions if only hourly traffic counts are available. The program can model the effects of flared minor roads and major road right-turning traffic blocking through traffic.

8.3   Roundabouts

The first roundabouts were constructed as 'circuses' with all vehicles entering the junction and turning left to merge with those circulating. There were no priority markings and vehicles were expected to 'zip' together by synchronising their speeds. Vehicles then circulated around the central island and diverged at their chosen exit. Two weaving manoeuvres were performed; one on entering the junction and the other just prior to the chosen exit. Under heavy flow conditions, roundabouts tended to 'lock', which occurred when drivers were prevented from leaving the junction by those entering. In the mid 1960s offside priority markings were introduced which reduced entry speeds, accidents and the tendency to 'lock'.

The offside priority rule radically altered the method of operation of roundabouts. Weaving formulae for roundabouts (Wardrop[5]) had produced very large junctions with wide straight links which encouraged high circulating speeds. Researchers at the Road Research Laboratory led by Mr FC Blackmore[6] experimented with very small central islands. Dramatic increases in capacity were achieved at smaller and smaller sites until the central island was reduced to a white painted circle. The entries were widened to allow for multi-lane approaches. The small size of the islands opened up many possibilities and many small priority junctions, including staggered junctions, were converted with significant reductions in accident rates. At several sites, old style roundabouts with long wide weaving sections were converted to multiple mini-roundabouts, with traffic travelling in both directions around the central island (e.g. Swindon and Hemel Hempstead) (Figure 8.9).

However, the smaller islands led to problems with speeding vehicles on nearly straight paths through the junction. Recommended designs now incorporate speed reduction measures that include:

- deflection islands;
- larger or domed central islands;
- nearside kerb buildouts.

The designs of large roundabouts were also affected by the Road Research Laboratory (now TRL) research and most roundabouts are designed with circular central islands and flared approaches. Further research led to the capacity formulae used by the *Assessment of Roundabout Capacity And DelaY* (ARCADY[7]) computer program.

The most important part of any roundabout design is ensuring that drivers approach it at suitable speeds. The DMRB Technical Standard TD 16/93[8] describes the entry curvature required to limit speeds but earlier advice on preventing a 'see through' alignment is worth considering.

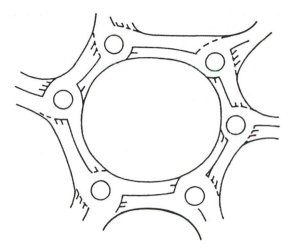

Fig. 8.9 Conversion of a conventional roundabout to a ring junction.

Speed reduction at small and mini-roundabouts can be difficult to achieve as space is limited and vehicle swept paths have to be accommodated. Many small roundabouts have been equipped with abrupt kerb deflections that only serve to complicate the appearance and often contribute very little to speed reduction and safety.

The vertical alignment is often under the control of highway engineers who are highly skilled at providing the correct drainage crossfalls on roads and junctions. The old style roundabouts were usually drained partly towards the central island and partly to the edge. If this method is adopted for small and mini-roundabouts the roundabout can be hidden from the driver's view. On larger roundabouts the superelevation around the central island can encourage excessive speeds. Central islands should be raised in the centre of the junction and the carriageway drained towards the edges. A good reference for the design of mini-roundabouts is 'Mini-Roundabouts – Getting them right!' by Sawers.[9]

8.3.1   ROUNDABOUT CAPACITY ASSESSMENT

The TRL Computer program ARCADY uses empirical formulae to model entry capacities and time-dependent queuing theory to estimate queues and delays for all types of single island roundabouts. The program requires the input of demand traffic and pedestrian flows (where pedestrian crossings are present) and various geometric parameters which describe each roundabout entry and the overall size of the junction. The program can simulate peak traffic conditions if only hourly traffic counts are available. The program produces consistent results for conventional and small roundabouts but when used for mini-roundabouts, it can underestimate potential capacity.

8.4   Signalled junctions

The general layout of signalled junctions is closely controlled by the proposed signal control method and the presence of pedestrian, cycle and bus facilities. It is normal to accept narrow, and sometimes sub-standard, lane widths at the junction stop lines. The advantage of multiple lanes often far outweighs the occasional problem with two or more lanes being dominated by a wide vehicle making a turn.

The requirements of pedestrians, cyclists and buses must be 'designed in' from the beginning of the design process. They cannot be ignored and then tacked on to a near complete design after detailed capacity assessments have been made.

In recent years the emphasis at traffic signals has moved away from pure handling capacity to provision of facilities for other users. In city centres pedestrian subways have lost favour and most new schemes include pedestrian crossings at ground level. Cycle advance areas are becoming more common and selective vehicle detection (SVD) for buses is in the process of becoming universal in many cities.

The increasing complexity requires careful attention to detail as there are more traffic islands being installed to accommodate ever increasing numbers of signal heads with pedestrian and cycle aspects (Plates 8.1–8.3). Large vehicle turning circles and swept paths must be rigorously checked. Very tight turns can slow large vehicles to the point where capacity is affected.

Plate 8.1 Cycle contra flow lane.

Plate 8.2 Cycling in traffic.

Plate 8.3 Off road cycle lane.

Carriageway and footway surfacing materials require careful consideration to complement the operation of the junction and to reduce the confusion to some road users. Tactile surfaces for pedestrians must be carefully placed to ensure that people with visual handicaps are helped safely through the junction. Traffic signal control is discussed more detail in Chapter 9.

References

1. Highways Agency, DMRB Volume 6 Section 1 Part 1 *Highway Link Design*, TD 9/93 – Amendment No1, HMSO, London.
2. Department of Transport (1988) *TRACK prediction of vehicle swept paths*, Savoy Software.
3. Highways Agency, DMRB Volume 6 Section 2 Part 6 TD 42/95 *Road Geometry Junctions, Geometric Design Of Major/Minor Priority Junctions*, HMSO, London.
4. Department of Transport (1985) *PICADY2: An enhanced program to model capacities queues and delays at major/minor priority junctions*, TRRL Research Report 36, TRRL, Crowthorne.
5. Wardrop, JG (1957) The Traffic Capacity of Weaving Sections of Roundabouts, *Proceedings of the First International Conference on Operational Research* (Oxford, 1957), pp 266–280, English Universities Press Ltd, London.
6. Ministry of Transport (1970) *Capacity of single-level intersections*, RRL Report LR356, TRRL, Crowthorne.
7. Department of Transport (1985) *ARCADY2: An enhanced program to model capacities queues and delays at roundabouts*, TRRL Research Report 35, TRRL Crowthorne.
8. Highways Agency, DMRB Volume 6 Section 2 Part 3, TD 16/93, *The Geometric Design of Roundabouts*, HMSO, London.
9. Sawers, C (1996) *Mini-roundabouts – Getting Them Right!*, Euro-Marketing Communications, Canterbury.

9
Signal Control

9.1 Introduction

Traffic signals are used to regulate and control conflicts between opposing vehicular or pedestrian traffic movements. Without the use of signals at some sites the major flow would dominate the junction, making entries from the minor road impossible or very dangerous. At other sites the minor road might interfere with the flow of major road traffic to such an extent that excessive congestion would occur. Traffic signals cannot only improve junction capacity, but can also improve road safety.

Modern traffic signal controllers utilising microprocessors are reliable and flexible and can be programmed to handle multiple phases and other features, including:

- pedestrian facilities;
- pre-determined fixed time;
- vehicle actuation including 'hurry calls' actuated by excessive queues on certain arms, public transport or emergency vehicles;
- links by cables or cableless linking facilities to other nearby signalled junctions or pedestrian crossings and integration into an urban traffic control (UTC) system, including adaptive systems such as *s*plit *c*ycle time and *o*ffset *o*ptimisation *t*echnique (SCOOT).

9.2 Fixed-time control

In fixed-timed control the proportion of green time assigned to opposing arms is pre-set in accordance with 'historic' traffic data. At most sites, traffic flows vary throughout the day. Typically inbound flows to a city are high in the morning peak period and outbound flows are high in the evening. Different green splits may, therefore, be required at different times of the day or year.

9.2.1 VEHICLE ACTUATION

Information about traffic demands on the approaches to signals is detected by inductive loop detectors buried in the road surface or by post-mounted micro-wave detectors. The signal controller can extend the green stage on the relevant arms from a preset minimum to a maximum. These systems are highly responsive and can minimise delays and maximise capacity at isolated independent sites. 'Hurry calls' to prevent excess queuing or to assist emergency vehicles and buses are very effective. Time savings for buses fitted with transponders at selective vehicle detection (SVD) sites can be as much as 10 seconds per bus.

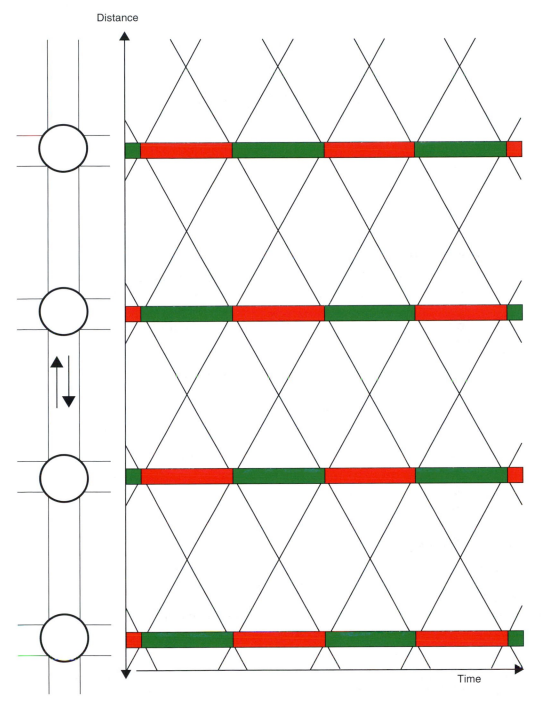

Fig. 9.1 Time and distance diagram.

Neighbouring traffic signals can be connected by cable links to provide uninterrupted progression for vehicles on the main route. The difference between the start of the green stage and the downstream junction is related to the journey time at cruise speed. The difference in the start of the green stages is known as the offset. Successive sets of signals can also be linked to produce a steady progression along a road. Signal offsets can be calculated or arrived at graphically, using a time and distance diagram (Figure 9.1).

Modern microprocessor controllers are equipped with digital time clocks which can be very accurately set using hand-held terminals. This facility allows signal timings and offsets to be synchronised without the need for a physical correction. However, interrupted power supplies or failure of a component can cause synchronisation to be disrupted.

9.3 UTC

An extension of the linking and co-ordination of adjacent signals to a whole network of signals can reduce journey times and congestion by maximising the network's capacity. Networks of signals are much more complex than simple linked signals as there may be more than one major route through the network and these routes may not be easy to identify. As pre-determination of effects is difficult to achieve, the use of a computer model of the network is needed to optimise the signal settings for all junctions (e.g. TRAffic Network StudY Tool, TRANSYT[1]).

UTC systems produce many benefits by increasing the capacity of the whole network and reducing travel times, fuel consumption and hence, air pollution and noise. Unfortunately this additional capacity is quickly filled by additional traffic and, apart from the steadier progression through the network, similar levels of congestion return.

Clearly any additional capacity created by UTC systems must be used wisely and that a strategy for its use must be agreed before implementation. Spare capacity can be utilised to provide bus priority measures, cycle facilities, pedestrian crossings and to alleviate 'rat-running' through residential and shopping streets. The Transport Research Laboratory (TRL) has estimated that a well-maintained fixed-time UTC system using TRANSYT plans can produce journey-time benefits of around 15% and a SCOOT[2] fully adaptive system of around 20%. Adaptive systems, such as SCOOT, use real-time data from on-street detectors and a computer model to continuously update signal timings.

9.3.1 FIXED-TIME UTC

Large UTC systems are controlled by a central computer which initiates signal plans for varying flow patterns. The plans are routinely activated by a clock within the controlling computer but can be overridden manually; or special plans can be programmed in to accommodate unusual or special events such as sports and entertainment. The greatest problem with fixed-time UTC systems is that the plans need to be updated regularly to keep pace with changes in traffic pattern. This represents a considerable expense in organising and collecting traffic survey information and the heavy workload often results in the work being given a low priority.

9.3.2 ADAPTIVE, TRAFFIC RESPONSIVE UTC SYSTEMS

Adaptive systems require three main components:

- vehicle detectors;
- central controlling computer;
- implementation of signal settings within the traffic signal controller.

Data are transmitted to and from the street over public telephone lines, purpose-built communication systems or, occasionally, radio links.

9.3.3   PLAN SELECTION

Traffic flow data are used to select appropriate, pre-determined fixed-time signal plans. This is similar to fixed-time UTC systems that use the time of day for plan selection. The fixed-time plan chosen will always be a compromise and each plan must be prepared using a tool such as TRANSYT. The Sydney Co-ordinated Adaptive Traffic System (SCATS) developed in Australia, for use in Sydney, is a successful example of this type of system.

9.3.4   FULLY ADAPTIVE SYSTEMS

Fully adaptive systems use traffic flow data to calculate optimum signal settings continuously and adjust signal settings in a controlled manner. Early adaptive systems suffered from a number of problems such as slow response, inadequate prediction and too-frequent plan changing.

The most widely used fully adaptive system in the UK is SCOOT.

SCOOT was the result of co-operation between the three main traffic signal supply companies in the UK, the Transport and Road Research Laboratory (TRRL, now TRL) and the Department of Transport.

SCOOT uses vehicle detectors (usually inductive loop detectors cut into the road surface) placed at some distance from the signal stop line. Correct siting of detectors is critical to the success of a SCOOT system. The detectors should be placed as far as possible from the stop line. There should be minimal change in traffic flow between the loop and the stop line and should be at least 10–15 metres downstream of the preceding upstream junction. Figures 9.2–9.4 show typical SCOOT loop sites and dimensions.

There should be at least one loop installed for each signal stage. Additional loops can be used for right-turn stages and for major flow sources. For example, a car park exit between the main loop and stop line could be provided with a loop (Figure 9.5).

Loop occupancy is checked four times per second and is used to provide information on traffic demand, queues, congestion and exit blocking. The unit of flow and occupancy used for SCOOT loop detectors is the link profile unit (LPU). For comparison purposes there are 17–18 LPUs/vehicle (e.g. a saturation flow of 2000 vehicles/hour is equivalent to a SCOOT saturation occupancy (SatOcc) of 10 LPUs/second, a 10 vehicle queue equates to 170–180 LPUs).

Data from the detectors is transmitted via an outstation transmission unit (OTU) to an instation transmission unit (ITU) in the SCOOT control room.

The SCOOT computer runs an on-line traffic model of the network or networks, under its control, a database containing information about the junctions being controlled, the loop detectors and their stop line distances and relationship to the junction. The traffic model calculates the optimum signal settings and then transmits the new timings to the individual signal controllers on-street.

The optimisers within the SCOOT model produce a large number of very small changes. Large changes can be very disruptive to traffic flow and can confuse drivers.

9.3.5   SPLIT OPTIMISER

The aim of the split optimiser is to minimise delay by minimising the degree of saturation and congestion. Each junction is optimised in turn, 5 seconds before every stage change. A maximum

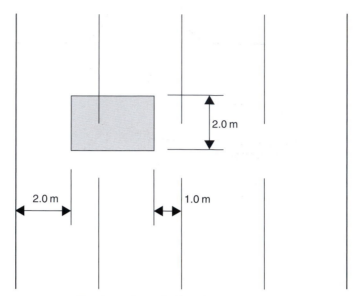

Two lanes in each direction

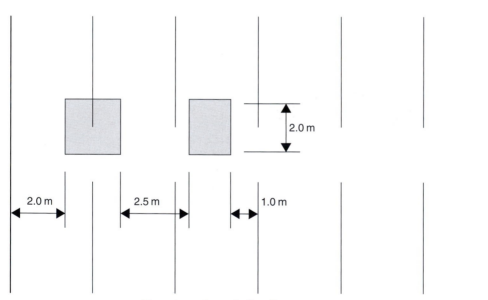

Three lanes in each direction

Fig. 9.2 Loop detector positioning for SCOOT.

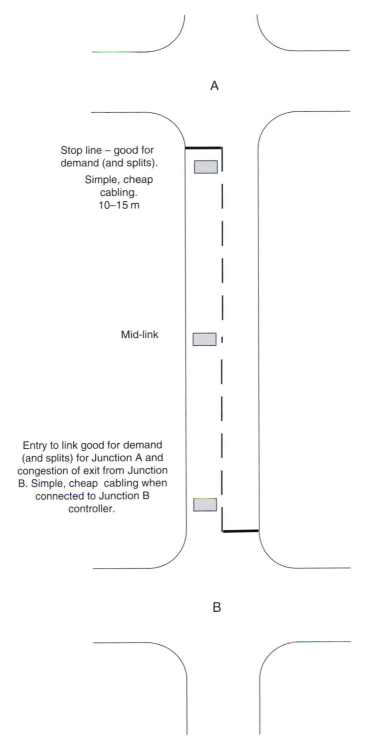

Fig. 9.3 Loop positions along a link.

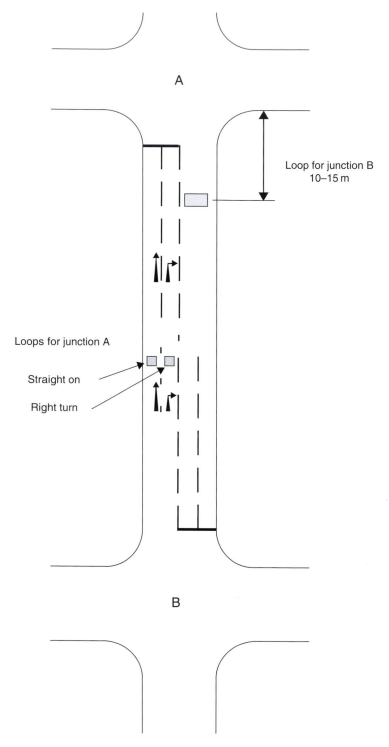

Fig. 9.4 Loop detector locations.

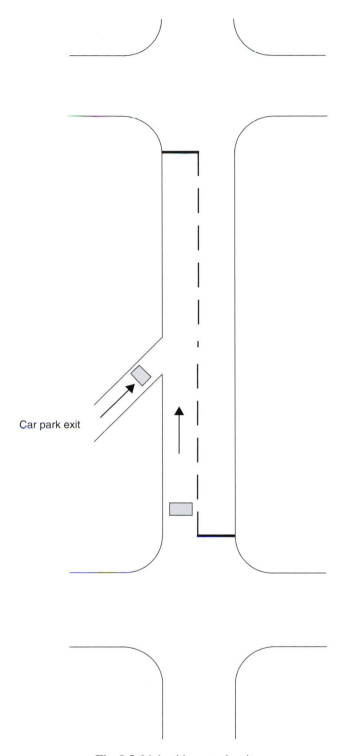

Fig. 9.5 Link with car park exit.

of 4 seconds temporary advance or retard is permitted with a permanent 1 second change within the pre-determined maximum and minimum green times.

9.3.6   OFFSET OPTIMISER

The offset optimiser aims to smoothen the flow through successive junctions. All upstream and downstream links for every junction are optimised for every cycle. Where links are very short, they can be equipped with fixed offsets; offsets can also be weighted or biased for specific purposes such as bus priority.

9.3.7   CYCLE TIME OPTIMISER

The objective of the cycle time optimiser (CTO) is to achieve 90% target saturation. Cycle time is adjusted every 2½ or 5 minutes between a preset minimum practical and maximum range in 4 second steps. The whole network runs on the same cycle time of half-cycle time for certain double cycle time systems and pedestrian crossings (Plate 9.1).

9.3.8   SVD

SVD has the potential to reduce bus delays at independent junctions by 32% (22% in UTC networks) and reduce variability by 22%. SVD uses detectors to advance or extend traffic signal stages to provide priority for buses. Buses are detected by many different methods including: loops cut into the road surface and bus-mounted transponders, beacons mounted on roadside street furniture and by global positioning satellite (GPS) systems. The efficiency of SVD is dependent upon the precise location of the vehicles, in relation to the junction, and the timely transmission of a vehicle's presence to the signal controller or the UTC computer.

9.3.9   SVD DETECTOR LOCATION

To obtain effective bus priority within the SCOOT-UTC areas, all approaches to signalled junctions with bus services will require detectors. Ideally, detectors should be sited downstream of the nearest bus stop and, normally, 70 metres from the signal stop line. However, useful benefits for buses can be obtained if this distance is no less than 35 metres and limited benefits can still be obtained below 35 metres. On higher speed roads and in free-flow conditions the optimal distance may increase up to 120 metres.

Plate 9.1 Cycle advance at traffic signals.

Transmission of the position of the bus must be arranged so that the bus priority sequence is not activated before it is needed. Therefore, the precise location should take account of the normal stopping place of buses. Figure 9.6[3] shows where carriageway loops and roadside detectors should be placed. A GPS system would use on bus transmission equipment to request a priority sequence when the bus reaches a similar point.

9.3.10   SCOOT VALIDATION

The SCOOT model uses data from its detectors to optimise signal settings and the resulting queue lengths. To ensure that the model produces the best possible signal settings, it is necessary to tune or calibrate the model so that it reflects the actual traffic conditions, as closely as possible. Once validated there should be no need to re-validate unless there is a major change in site conditions or traffic composition. The following changes at a site would create a need to re-validate:

- loop resiting;
- change in junction geometry;
- addition or deletion of traffic flow source;
- change in speed or parking regulation;
- change in composition of traffic, vehicle size or performance;
- technological changes to vehicles or control systems.

Validation requires a team of two people on site, usually a traffic engineer and a technician, and a two-way voice communicator with the control centre. It is also possible to use a hand-held terminal attached to the OTU in the signal controller to monitor SCOOT input/output messages. Approximately 1 day is required to validate a typical junction.

9.3.11   SCOOT VALIDATION PARAMETERS

- *Journey time*: the time taken by an average vehicle in a free-flowing platoon between the detector and the stop line.
- *SatOcc*: maximum outflow rate of a queue over the stop line in LPUs/seconds:

  SatOcc (LPUs/seconds) $\times$ 200 = saturation flow in vehicle/hour approximately.

- *Maximum queue*: average largest number of vehicles that can normally be accommodated between the detectors and the stop line.
- *Green start lag*: inter-green plus 1 or 2 seconds.
- *Green end lag*: 2 or 3 seconds.
- *Main downstream link*: determined from local knowledge and observation.
- *Default offset*: journey time minus standing queue clearance time.

9.3.12   SCOOT DATA

The SCOOT model requires input data to be held in a structured database. The SCOOT hierarchy is as follows:

- *Area.*
- *Region*: part of an area for which there is a common cycle time.
- *Node*: individual signal-controlled junctions or pedestrian crossings.
- *Stage*: junction signal stage.

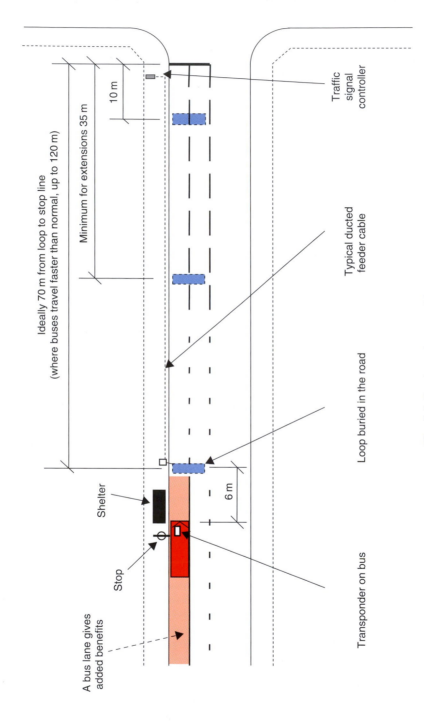

Fig. 9.6 Bus loop detector siting.

Ideally 70 m from loop to stop line
(where buses travel faster than normal, up to 120 m)

Minimum for extensions 35 m

10 m

Traffic signal controller

Typical ducted feeder cable

Loop buried in the road

6 m

Transponder on bus

Shelter

Stop

A bus lane gives added benefits

- *Link*: normal, entry, exit, filter on bus priority link.
- *Detector*: detector identity – there may be more than one detector on a link and these should be given separate identity.

SCOOT can be used to favour certain links or series of links but overall network delay will increase. Occasionally users of a SCOOT network observe that a particular junction is not operating at capacity. It should be remembered that SCOOT is optimising a whole network and that the individual junction is an integral part of that network. It is possible that, if the junction is altered in isolation, queues will build up on another part of the network resulting in an increase in overall network delay.

Automatic *SCOOT TR*affic *I*nformation *D*atabase (ASTRID) is an associated program which can be used to extract historic data from the SCOOT database. This information is very accurate and can be obtained for extended periods. Data is supplied in LPUs and then converted to vehicles using the standard 17–18 LPUs/vehicle or calibrated by conventional traffic counts.

9.4   Other UTC facilities

9.4.1   FAULT MONITORING

Certain fault conditions can be transmitted directly to the control centre and used to direct maintenance operations.

9.4.2   CAR PARK INFORMATION AND VARIABLE MESSAGE SIGN

Additional detection can be used to monitor car park arrivals and departures. Space availability is then displayed on variable message signs (VMS) installed at key locations.

9.4.3   'HURRY CALLS' AND 'GREEN WAVES'

Excess queues can be dissipated when queue detectors are occupied at important sites. Priority routeing for emergency vehicles can be implemented using pre-determined plans or by using transponders fitted to the vehicles and special detectors. Transponders are commonly fitted to buses for conventional SVD junctions and more recently in fixed-time UTC and SCOOT areas.

9.5   Traffic signal capacity assessments

9.5.1   INDEPENDENT/ISOLATED JUNCTIONS

In recent years there has been a rapid increase in the number of traffic signalled junctions, either linked to adjacent junctions or under UTC. However, the majority of signalled junctions operate independently of others. This has the advantage that cycle times and green splits can be optimised to a high level without the need for compromises demanded by nearby junction. The traffic engineer must be able to assess the likely queues and delays when changes are being made to existing junctions or new installations are being considered. The assessments can be divided into two parts:

- capacity, queues and delays;
- signal settings or timings.

The capacity of a junction is dependent on the green time available on each arm and the maximum flows over the signal stop lines (saturation flow).

For a single traffic stream during one cycle of the signals, the capacity is:

$$\mu = \lambda S$$

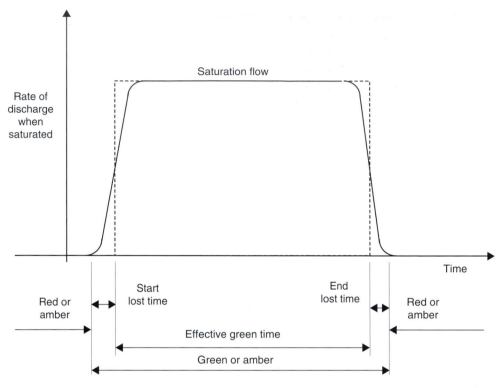

Fig. 9.7 Saturation flow profile.

where

μ = maximum flow rate over the stop line/unit time;
λ = proportion of the cycle which is effectively green;
S = saturation flow for the approach.

Traffic flow is expressed in passenger car units/hour (PCU/h). PCU/vehicle ratios for different vehicle types at traffic signals are as follows:[4]

Cars and light goods vehicles	1.0
Medium goods vehicles	1.5
Heavy goods vehicles	2.3
Buses and coaches	2.0
Motorcycles	0.4
Pedal cycles	0.2

Saturation flow occurs when there is continuous queuing on the approach and is determined, at existing junctions, by direct measurement or from empirical formulae derived from public road and track tests carried out by the TRL.[5-7]

These formulae predict saturation flows from measured geometric parameters at the junction as follows:

● lane width;
● gradient;

- lane position (i.e. nearside or non-nearside);
- vehicle type composition;
- turning radius (as appropriate).

Effective green time is used to simplify the typical flow profile (Figure 9.7), which assumes that there is a queue on the approach before and after the end of the green stage.

The practical capacity of an approach or stream is:

$$\frac{psg}{c}$$

where

g = effective green time;
c = cycle time;
p = maximum practical degree of saturation is usually taken as 0.9;
s = saturation flow.

All approaches should operate within their practical capacities or the junction will be over-loaded. If the demand flow is less than the practical capacity, the differences are termed the reserve capacity.

There are a number of computer programs available for the assessment of independent signalled junctions. The most popular in current use are Optimisation of Signal Capacity And DelaY (OSCADY)[4] and LINSIG.[8]

OSCADY was developed by the TRL optimises signal timings for a period of the day with varying demand flows and calculates the resulting queues and delays. Queues and delays are calculated using time-dependent queuing theory similar to that used in the Assessment of Roundabout Capacity And DelaY (ARCADY) and Priority Intersection Capacity And DelaY (PICADY) programs.[9] OSCADY also produces signal settings but LINSIG has gained popularity with traffic signal engineers in recent years.

TRANSYT was developed by TRL to optimise timings for a network of traffic signals. The resulting timings are used for fixed-time UTC systems and to provide initial settings for SCOOT.

TRANSYT models the dispersal of platoons of vehicles leaving upstream junctions as they travel towards the next junction and automatically selects the best timings for the network. Priority junctions and roundabouts can be modelled reasonably well if they occur within the network but the program is intended primarily for fully signalled networks. The network is represented by a series of nodes (junctions) and links (traffic streams). A single link can represent several traffic streams at a stop line or several links can share the same stop line. Stop line saturation flows, link lengths, link cruise speeds and demand flows are supplied to the model. The model calculates a performance index (PI) based on the monetary value of delays and stops for each set of trial signal settings. The optimiser searches for the minimum PI for the network.

References

1. Department of the Environment (1980) *User guide to TRANSYT version 8'*, TRRL Laboratory Report 888, TRRL, Crowthorne.
2. Department of Transport (1981) *SCOOT: a traffic responsive method of coordinating signals*, TRRL Laboratory Report 1014, TRRL, Crowthorne.
3. Traffic Control Systems Unit (1997) *Bus priority – selective vehicle detection in London*, (unpublished, may be available on request), London Transport Buses, London.

4. Department of Transport (1987) *OSCADY: A computer program to model capacities queues and delays at isolated traffic signal junctions*, TRRL Research Report 105, TRRL, Crowthorne.
5. Department of Transport (1986) *The prediction of saturation flows for road junctions controlled by traffic signals*, TRRL Research Report 67, TRRL, Crowthorne.
6. Department of Transport (1982) *Traffic signalled junctions: a track appraisal of conventional and novel designs*, TRRL Laboratory Report 1063, TRRL, Crowthorne.
7. Webster, FV and Cobbe, FM (1966) *Traffic signals*, Road Technical paper 56, HMSO, London.
8. Simmonite, B (1985) LINSIG: A program to assist traffic signal design and assessment, *Traffic Engineering and Control*, **26**(6).
9. Kimber, RM and Hollis, EM (1979) *Traffic queues and delays at road junctions*, Department of Transport, TRRL Report LR 909, TRRL, Crowthorne.

10
Parking: Design and Control

10.1 Introduction

The traffic engineer will need to know how best to provide parking, and how to control parking facilities, both on- and off-street, both in surface sites and structures. In this chapter we describe the key factors that need to be considered. We also talk briefly about the need to ensure that any off-street car park is maintained in an appropriate state.

Parking provision on the highway in Great Britain is constrained by legislation. Government rules and guidelines determine where parking can be provided, the methods of control and the design standards to be used. Separate legislation applies in Northern Ireland.

Off-street car parks are provided to meet a variety of needs and the type of need can affect the design of the car park. It is widely accepted, although one has to admit without any hard evidence, that drivers will have fewer problems using lower standard car park that they are familiar with, and so car parks used by the same parkers day after day can be designed to a lower standard than one would otherwise expect. For example, it is often argued that an office car park could be designed to a lower standard than would be considered necessary for a public car park, as most of the users will be familiar with the geometry of the car park. Also if, for example, cars are parked too close together, a work colleague can usually be found and asked to move his/her vehicle with reasonable ease.

Although off-street car park design is not governed by legislation in the same way as parking on the highway, the operation of a public car park can be subject to regulation using statutory powers. Further, with the growth in concerns about safety in car parks there are new constraints that affect the design and aftercare of car parking structures.

Car parking control equipment is becoming increasingly more sophisticated, and the advent of microprocessor-based systems has allowed parking control systems to become more flexible, to meet the varied demands of users more closely. However, just as the availability of increasingly sophisticated control systems has affected the way parking is controlled off-street, on-street parking controls have seen a different kind of revolution. This has allowed more flexibility, to better meet users' needs, through the use of both high-tech and very low-tech control equipments.

10.2 On-street parking

Most roads, in most places, are not subject to any form of parking control; and it is widely believed that there is a right to park where no controls are present. This is not true. There is a right to pass along (travel) on a highway but no absolute right to stop. Indeed as far back as 1635, parking problems in London had become so bad that a Royal proclamation was issued which declared that:

> We expressly command and forbid, that no hackney or hired coach be used or suffered in London, Westminster or the suburbs thereof, …

Earlier still in ancient Rome, things reached such a state that in an attempt to restore order the death penalty was threatened to anyone leaving a chariot parking on a public thoroughfare. There are many who, faced with/dealing with today's traffic and parking problems look back wistfully to the enforcement mechanisms available in earlier times!

Generally speaking, a vehicle parked at the kerbside, providing it is not parked dangerously or blocking traffic, will not attract police action although, in theory, the driver could be prosecuted for causing a highway obstruction.

Formal parking on the highway can be provided for general use, or for particular groups of users. For example, kerbside parking could be set aside for a particular class of vehicles; this could include:

- all vehicles below a certain size (i.e. excluding heavy commercial vehicles)
- solo motorcycles
- taxis
- buses
- vehicles belonging to registered disabled
- local residents' vehicles
- local businesses
- doctors
- diplomats
- police vehicles or
- specialist vehicles, such as a mobile library.

The provision can be made all the time (at any time), on certain days or at certain times of day, and the provision can be free or be charged for. Thus, generally speaking, a taxi stand is available only to taxis at any time. However, other facilities may only operate part time.

For example, parking for residents in a city centre may only be protected during the working day, typically 08.30–18.30 hours, Monday–Friday, although the increasing tendency for late night opening and the introduction of Sunday trading means that the timing of many of these regulations will increasingly be rethought to be 24-hour restrictions.

Where activity is intermittent, for example, near a football stadium, restrictions may not be appropriate all the time and may only apply on match days. This type of restriction is increasingly common in areas where a local community requires occasional protection from a short-lived sudden influx of parkers associated with activity such as a sports stadium.

10.2.1   ROAD MARKINGS

The design and marking of parking bays on the highway is governed by legislation, with approved markings shown in Schedule 6 of the Traffic Signs Regulations and General Directions (TSRGD) 2002, SI 2002/3113.[1] These show a variety of markings for both cars and for specialist bays, for buses, taxis etc. (Figure 10.1). Echelon parking, as shown in diagram 1033, is seldom used, as, although it may allow additional parking spaces to be accommodated in a wide street, drivers tend to have difficulty manoeuvring in and, particularly, out of the spaces safely.

The current regulations specify a standard bay width, of between 1.8 and 2.7 metres, although a bay width of 2.7–3.6 metres is specified for parking bays for the disabled. In reality, these

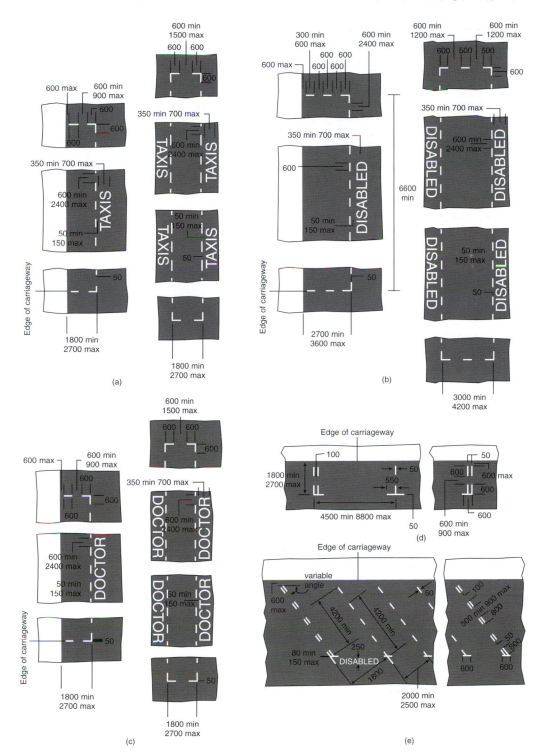

Fig. 10.1 Road markings for special types of parking.[1]

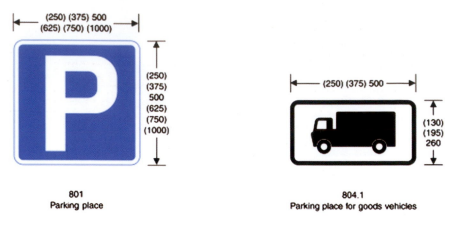

801
Parking place

804.1
Parking place for goods vehicles

Fig. 10.2 Parking signs for particular types of road users.[1]

widths may not be achievable in narrow streets, and it may be tempting to make a pragmatic trade-off between the need to provide parking and the need to ensure sufficient road width for moving traffic. However, the regulations are prescriptive and if a local authority were to mark narrow bays, they would be unenforceable. Prior to the earlier 1994 regulations, government guidance allowed bays as narrow as 1.6 metres. Below this, the width of the average modern car would mean that many vehicles would protrude outside the bay.

The markings signs and design rules shown in TSRGD 2002 are not subject to variation other than as set out is the statutory instrument and it must be understood that any sign or line which does not conform entirely to the regulations has no effect, unless prior special approval has been obtained from the Government.

10.2.2 SIGNING

Where parking is provided with restrictions then signs are used to show what type of vehicle can park and how. These regulatory signs are specified in Schedule 2 of the TSRGD 2002. The signs either specify the type of vehicle allowed to use the parking place, or specify the conditions and restrictions attached to the use of the parking place. If the parking is available for a particular class of vehicle to use, without other constraint, the signing will identify the class of vehicle that can use the parking (Figure 10.2). Alternatively, the signing may display information about the limits and conditions of use for parking. Examples of this type of parking signs are given in diagrams 660.5 and 661.1–4 (Figure 10.3).

10.2.3 REGULATIONS

Parking on the highway can be provided using a number of different methods of regulation. The principal legislation is the Road Traffic Regulation Act 1984 (RTRA 1984),[2] as modified by later legislation.

Part I of the Act provides general powers to allow traffic regulation. Using powers in this part of the Act, a highway authority can make regulations to restrict parking, or within the wording of the legislation, 'waiting and loading'. This legislation also allows for exemptions to be made to any order and so some parking provision to be made 'by exception'. This means that, in a street where parking is otherwise banned, a parking place may be provided in this way, for example for motorcycles. Thus

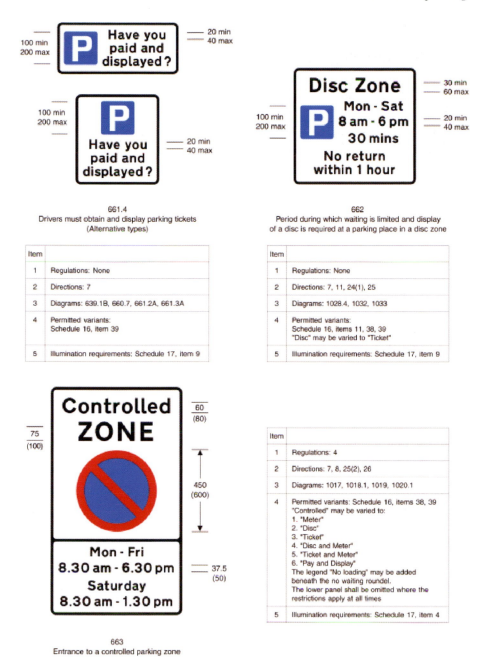

Fig. 10.3 Parking restriction signs.[1]

if any other vehicle parks in a motorcycle bay it is in breach of the more general order banning waiting. Great care has to be taken in using the law in this way. If a street had a waiting restriction, without a ban on loading, with, say, a motorcycle parking bay, then any other vehicle parking in the bay would be committing an offence but, if the vehicle were loading goods, then no offence would be committed! This is clearly confusing and unsatisfactory. 'By exception' provision is of necessity uncharged, since the exception is granted for the vehicle type, not against a payment.

Part IV of the Act provides powers to create parking places and can be used to create specific parking places on the highway. Using these powers, the highway authority can make explicit provision for parking in a street, either free or for payment. Part I powers would not be used, for example, in a street to provide, say, a disabled bay where no other controls were proposed. In these circumstances, the bay would be specified; using the powers contained in Part IV of the Act. There is a general presumption that vehicles with a blue badge will have exemption from most waiting and loading restrictions although special rules apply in parts of London, and on red routes.

In places where parking controls are needed on an area-wide basis, the required provision is made by creating a controlled parking zone (CPZ). Within a CPZ all the streets can be controlled by an overall order, which is applied to all the streets within the area. Normally this is a restriction on parking for part or all of the day. The powers in Part IV of the Act are then used to specify places where parking is allowed. Typically these powers would be used to specify parking for residents, plus parking for visitors to the area. The logic of a CPZ order is to have an overall order which applies to all the streets, with specified areas then set aside for parking. This has the advantage that, if a parking place is no longer needed, then the underlying restriction automatically applies as soon as the parking is removed.

The statutory responsibility and power introducing street parking controls resides with the highways authority. The procedures that a highway authority has to follow in seeking to introduce controls are set out in the Traffic Order Procedure Regulations and these are contained in two documents:

- Local Authorities' Traffic Order (Procedure) (England and Wales) Regulations 1989, Sl 1989/1120.[3]
- The Traffic Order Procedure (Amendment) Regulations 1993, SI 1993/1500.[4]

10.3   Off-street car parking

The design of a car park structure is a complex exercise requiring a detailed understanding of many factors. In this book we cannot hope to provide more than the basic issue that anyone embarking on this exercise needs to address.

The layout of a surface car park will be greatly influenced by the shape and form of the land over which it is constructed. Obviously a flat, square site offers an easier design problem than a steeply sloping irregularly shaped one. As a rule of thumb, to estimate the number of cars that can be accommodated on a site, each car space requires about 25 square metres of space.

The basic element in a car park is a rectangular parking space, which would typically be of the order of 4.8 metres × 2.4–5 metres. However, bays could be as small as 4.6 metres × 2.2 metres, where space is at a premium, and widths could be up to 3.6 metres or more, for disabled parking. Generally speaking, short-stay car parking, where there is a constant coming and going operates more efficiently with wide bays and the Institution of Highways and Transportation (IHT) recommend a 2.5 metres bay width for short-stay car parking.[5] Studies over time have shown that as car design has developed vehicle doors have tended to get thicker. A typical 1960s car would consist of little more than two sheets of metal with a window in between. By comparison the 21st century vehicle door also contains crash bars, airbags, electric motors speakers and storage bins. The same studies show that, overall cars are not getting significantly longer. However, with the increasing popularity of off-road four-wheel drive vehicles, headroom requirements are also increasing.

Parking bays are grouped together in rows and parking places at 90° tend to provide the most efficient use of a site, although, if the car park shape is irregular, echelon parking may be desirable to use the space more efficiently. Echelon parking may also be desirable in situations of very short-stay parking where dynamic capacity, being the maximum flow of traffic along aisles, is more important than static capacity.

The rows of parking have to be segregated by aisles to allow the vehicles to access the bays and the aisle width depends on whether it is carrying one- or two-way traffic. Opinions are mixed on the minimum required. A two-way aisle needs to be 6.95 metres according to the IHT, with a 6-metre aisle being acceptable for one-way traffic. Other sources recommend that a 6-metre aisle is sufficient to carry two-way traffic.

Although the lower figure may be defensible in strict functional terms, one should never forget that, in most car parks, one is providing a paid for service and therefore, the easier the car park is to use, the more attractive it will be to the payee and the more likely they are to pay a higher charge. Therefore, in designing a car park the engineer should look beyond the strict functional requirement to quality of service and ease of use.

Designing the physical layout of a car park should go beyond the engineering design of the parking spaces to include consideration of the aesthetic and environmental design of the car park, and the creation of a secure environment for cars and drivers. A system of car park design assessment has been created under the 'Safer Parking' scheme promoted jointly by the Association of Chief Police Officers and administered by the British Parking Association.[8]

The scheme seeks to identify safe car parking and considers factors such as:

- levels of lighting;
- landscaping and the extent to which landscaping offers places where criminals could hide, the guidelines recommends that planting should be below 1 metre or above 2 metres;
- levels of natural surveillance, that is whether the car park can be seen by people passing by;
- the level of security offered by CCTV or by patrolling staff.

There are a variety of designs for car parking structures. The floor of the car park can be flat with connecting ramps. Ramps connecting floors can be straight or can be helical.

The floor can itself be built on a gradient to obviate the need for connecting ramps, in effect creating a continuous spiral.

Where land is at a premium, mechanical parking systems can be used. These car parks typically use palletised systems to stack cars and require less space per car as there is no need to allow manoeuvring space. Mechanical systems have had limited appeal in the UK and this is probably because of their high capital cost relative to a traditional self-park car park, typically costing 2–3 times more per space than a traditional car park. Their value is in their ability to provide significant amounts of parking in small spaces where it would not be viable to build a traditional self-park car park.

10.4 Whole life care of car parks

Car park design and engineering standards are developing and evolving continuously, but historically effort has been focussed on the creation of the car park rather than maintaining it after it is opened. There are probably about 6000 car parking structures in the UK and many of these car parks were built in the 1960s and 1970s and have had little or no preventative maintenance from the day they opened. The consequences of this were dramatically demonstrated with the partial collapse of the Pipers Row car park in Wolverhampton in 1998.

Following this the government instituted research into the structural integrity of the nation's car parks by the Standing Committee On Structural Safety (SCOSS) and following the SCOSS report the Institution of Civil Engineers published 'Recommendations for the Inspection Maintenance and Management of Car Park Structures',[6] which set out a regime for the inspection and care of car park structures. Although the publication is the recommendation of a professional body, the Health and Safety Executive have indicated that, due to other legislation, the recommendations have the force of law in that were there to be an incident associated with a car park failure and the owner had not followed the recommendations they could be criminally liable.

10.5   Disabled drivers

Parking facilities designed specifically for, or to include, disabled drivers should be fully accessible throughout for wheelchair users. Parking bays should be 1.5 times the width of a normal bay with access by ramps with a maximum gradient of 5%. Where lifts are provided they should have controls at wheelchair level, that is at about 1.2–1.4 metres above ground level. In providing parking one should be aware of the provisions of the Disability Discrimination Act 1995 (DDA1995).[7] In simple terms, the Act, which comes fully into effect in October 2004, makes it illegal for a service provider (e.g. a car park operator) to discriminate against someone on the grounds of a disability.

This does not mean that all car parking has to be fully accessible to all disabled people; a provider may meet their obligation by making 'equivalent' provision. Thus a car park with stairs could have disabled parking reserved on the ground floor. The law will rely on case law for its interpretation as to what is reasonable and this is particularly relevant to the provision of parking. For example, parking provision in a town centre may be shared between the district council and private operators and it is a moot point whether the owner of a private car park would be required or expected to provide disabled parking facilities in their car park if the street outside already met this need.

It is a common and incorrect belief that when providing parking for people with disabilities the parking must be free. This is not the case; pragmatically, when the blue badge holder can park free on the street outside the car park there would appear to be little point in attempting to levy a charge, but the thrust of the 1995 Act is to ensure accessibility rather than to exclude charges.

10.6   Servicing bays and lorry parking

Shops and factories require service bays, to accommodate commercial vehicles delivering and collecting goods. The demand for service bays should be determined as part of a traffic/transport impact study, undertaken to assess the transport needs of a new development.

A loading bay would typically be 3.3 metres wide with a length determined by the scale of vehicle expected to use the bay, which could be up to 15.5 metres long. The access route to the loading bay should be sufficient to allow the commercial vehicle to access the bay and manoeuvre into and out of the bay with ease. A 15-metre articulated lorry typically has a 13-metre outer radius and a 5.3-metre inner turning circle and the design of the manoeuvring area for service vehicles should ensure that this swept path is clear of obstructions.

This must be approached with common sense. Lorries are driven by human beings and allowance must be made for the variability that this will introduce to the path that a lorry follows when manoeuvring. Thus it is not good practice, having determined the path of the vehicle to place an obstruction within a few centimetres of the theoretical limits of the lorries swept path. It is surprising how often otherwise sensible architects and engineers do just this.

10.7 Parking control systems

Parking on the highway can be controlled in a number of ways by regulation. The regulation can define:

- the days, or times of day, that vehicles can park;
- the length of time that a vehicle can park;
- the type of vehicle that can park;
- the charge for parking.

The responsibility for this regulation rests with the highway authority, which has to make a Traffic Regulation Order setting out the form of control required. The primary powers which enable the highway authority to control parking are set out in RTRA 1984, which has subsequently been amended by various enactments and statutory instruments. The main legislation is summarised in Table 10.1.

Table 10.1 Summary of main legislation in RTRA 1984

Enactment	Section	Application
RTRA 1984	1	Prohibiting waiting and loading outside London
	6	Prohibiting waiting and loading in London
	9	Making experimental orders
	32–42	To provide free parking on-street and off-street parking places
	43–44	Local authority powers to license public off-street parking places
	45–56	Parking on the highway for payment
	61	Loading areas
	63	Cycle parking
	95–121	Offences
	122	Exercise of powers
Local Government Act 1985		This legislation abolished the Greater London Council (GLC) and the other Metropolitan councils and amended the 1984 Act to pass the relevant powers to the successor bodies
Road Traffic Regulation (Parking) Act 1986		This Act amended the 1984 Act to allow the use of cashless payment systems
Road Traffic Act 1991	41	Simplification of procedures for changing off-street parking charges
	42	Simplification of procedures for changing on-street parking charges
	43	The power to create permitted and special parking areas outside London
	44	Parking attendants
	63–82	Decriminalised parking powers for London

661.1
Restrictions on length of
waiting time and return period

Fig. 10.4 Free parking restriction sign.

10.7.1   *FREE PARKING*

As car use and hence parking demand grows, the need to control parking increases. The simplest form of control is a system which seeks to limit how long a vehicle can park. In Great Britain this can be done using a regulation made under Section 35 of RTRA 1984 to set a limit on the length of stay. This regulation is signed using the sign shown at Figure 10.4, which requires that the vehicle once parked cannot return within a certain period of time. Enforcement of this system relies on parking enforcement staff recording the details of each vehicle they see parked and returning sufficiently frequently to identify vehicles that have stayed beyond the maximum stay, but have not had time to leave and legally return.

Thus, if the limit was 1 hour, with no return for an hour, the enforcement staff need to visit more frequently than once every 2 hours and less frequently than once an hour. This would ensure that, if they saw a vehicle parked in the same place on successive visits, it must have committed an offence as could not have left and returned after being away for the required hour.

Clearly, to ensure that this type of parking control works it is necessary to have either a high level of compliance, or a high level of enforcement. For the enforcement to be effective the patrol staff have to keep good records, making the system labour intensive and hence costly to run.

In some places where free parking is still available parking is controlled using a parking disc (Figure 10.5). The disc, which is described in BS 6571 Part 8[9] is a simple cardboard clock which requires the driver to set the time of arrival. This simplifies the enforcement process as now all the enforcement officer has to do is read the clock face to see if the vehicle has over-stayed. The driver commits an offence by over-staying or by not having a disc. Discs are also used by disabled drivers where their stay is time limited.

There are many towns where parking is still free, using one of these systems. However, as car ownership and use increases, local councils are tending to move towards systems where a driver has to pay for parking.

Fig. 10.5 Disc zone parking restriction sign.

10.7.2 CHARGED FOR PARKING

The reasons for introducing parking charges can be complex and include:

- the need to deter parkers;
- the need to cover the costs associated with providing the parking;
- wider transportation policy objectives.

Whatever the reason, the decision to introduce charges will mean that the highway authority will have to make a traffic order, to bring into effect the charges and introduce a new form of control in order to collect the charges.

As part of the decision to introduce charges, the authority may also decide to allocate parking space to certain groups, with spaces set aside for particular groups of users. Thus a council could set aside some spaces for local residents, some for local businesses, some for motorcycles and some spaces for the use of disabled drivers, with other spaces for all-comers.

Charges for on-street parking can be controlled by a number of ways. The most common ones are:

- clockwork parking meters;
- electronic parking meters;
- pay and display or
- parking vouchers.

A parking meter is used to control an individual parking bay, and allow a driver to pay for their parking at the time of parking. Clockwork parking meters generally conform to BS 6571 Part 1,[10] a typical example is shown in Figure 10.6. The operation is simple, the driver inserts coins and the meter registers the time paid for. When the time paid for has expired, the meter shows a flag in the display to show to a parking attendant or traffic warden that a penalty has been incurred.

In electronic meters, described in BS 6571 Part 2,[11] the clockwork mechanism has been replaced by an electronic timer, display and coin validation system. In addition to offering the greater accuracy and reliability inherent in an electronic mechanism, the electronic coin valid-ation system allows better checking of coins and for a greater variety of coins to be used than would be practical with a purely mechanical device. Figure 10.7 shows a typical example of an

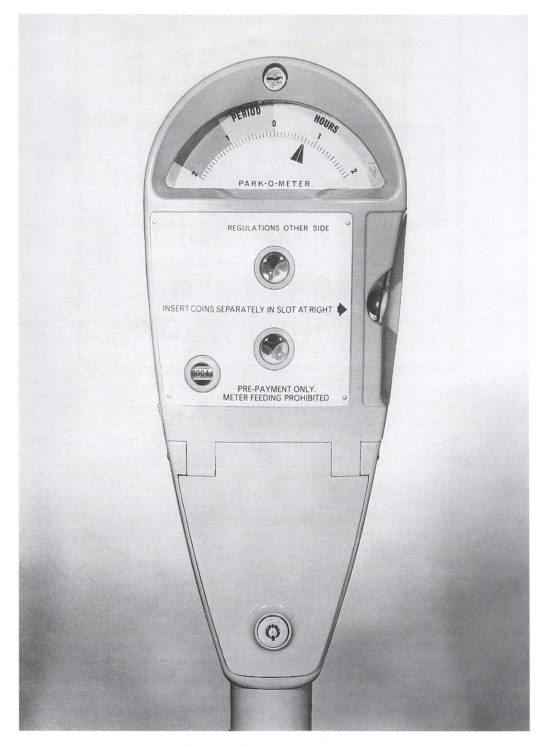

Fig. 10.6 Clockwork parking meter.

Fig. 10.7 Electronic parking meter.

electronic parking meter. Clockwork parking meters were the first system of automated parking payment available, but over time have become less popular because of their relatively high operational costs compared with pay and display machines. Some meters are still in use, particularly in locations where only one or two parking places can be fitted into a street. However, as equipment and expectations have become more sophisticated, few entirely new installations have been made.

A pay and display machine allows a large number of spaces to be controlled using a single machine. The number of bays controlled will depend on the layout of the streets and the closeness of the bays to each other. Like a parking meter, the driver has to pay for parking on arrival. However, since the machine controls a number of bays, the parker will have to walk to the machine to pay. The pay and display machine issues a ticket which the driver displays within the car, so that it can be checked by the patrol staff. Figure 10.8 shows a pay and display parking meter conforming to BS 6571 Part 3.[12]

Pay and display machines are electronic and can be powered by either batteries or from the mains. The machines allow, among other things, full audit systems, to monitor the payments being made. In addition, most machines now allow payment using a cashless system, usually a stored value card the size of a credit card. The sophistication of the equipment means that many machines have the capability for remote monitoring of their operation, either by radio or using a telecom line. Remote monitoring allows the operator to be warned if, for example, the machine develops a fault, is running low on tickets or has been broken into.

Some local authorities have introduced the option of paying for parking using standard bank credit/debit cards. The card is inserted in the machine and read and validated and the roadside and

Fig. 10.8 Pay and display parking meter.

the appropriate parking fee is charged to the card account. Given the relatively small charges involved these transactions are usually 'unauthorised' that is the card is checked against a local hot list but there is no on-line validation, such as takes place in a store. The recent launch of the new 'Chip + Pin' credit card where the user is normally expected to enter a pin code to validate their transaction will require pay and display machines to have a pin pad if this facility is to be used.

Although pay and display machines are considerably more expensive than a parking meter, their relative sophistication and lower operating costs have made them the more attractive option.

Parking vouchers are the most simple form of paid parking system. The voucher, (Figure 10.9) for an example, has to be pre-purchased, usually from a local shop. The parker scratches off the panels which relate to the time of parking and displays the card in the car's window. This can then be checked by patrol staff, to ensure that the time paid for has not been exceeded. Parking voucher design is specified in BS 6571 Part 7.[13]

A recent new initiative has been the introduction of pay by mobile phone systems, where a mobile phone is used to make a parking payment. There are a variety of systems on offer which

Fig. 10.9 Parking voucher.

can be categorised as:

- Systems where the mobile phone is used to interact with an existing pay and display unit to print a ticket rather than using coins or other payment media. These normally require the user to pre-register their use.
- Systems where the phone is used to communicate with a remote bureau to register a payment. These also require pre-registration.
- Systems where a payment is made using a text message. The most common system does not require pre-registration.

These systems are in their infancy and it is too early to judge what their long-term impacts will be. In most places these systems are offered as complementary services in addition to a pre-existing pay and display or voucher parking. However one system, in Belgrade, has been implemented as the primary means of payment, where although pay and display machines are also provided there are relatively few assuming that they will provide the back up to the primary, mobile phone system.

10.7.3   OFF-STREET PARKING

Off-street parking can also be controlled using a Traffic Order. The most common form of control is to use pay and display machines. This has the advantage of being well understood by the public and relatively cheap.

Other types of parking control can also be used for off-street parking. The main methods of control are summarised in Table 10.2.

Table 10.2 Other types of control for off-street parking

System	Description	Features
Manual, pay on entry	A driver pays an attendant on entry to the car park	Very simple, only allows a flat fee; weak accounting of income
Manual, pay on exit	A driver pays an attendant before exiting the car park	Simple, only allows a flat fee unless a time stamped ticket is issued on arrival, weak accounting of income
Automatic, pay on entry	Payment at a barrier on arrival	Only possible to use a flat fee, better accounting, requires either pre-paid card or season ticket or cash; relatively low lane throughput (*circa* 200 vph)
Automatic, pay on exit	Payment at barrier on exit	Only possible to use a flat fee, unless linked to an entry barrier which issues a time stamped ticket; better accounting; requires either pre-paid card or season ticket or cash; relatively low lane throughput (*circa* 200 vph)
Pay on foot	The driver takes a time stamped ticket on entry to to the car park; on return, the driver pays on foot for parking at a pay station and then takes an exit ticket which allows exit from the car park	Highly sophisticated systems allowing payment away from the entry/exit lanes, using coins, banknotes, or credit cards; by separating the payment action from access the throughput at a barrier can be increased to over 400 vph

The most sophisticated pay on foot systems are developing links with CCTV whereby rather than issue the user with a ticket the system reads and stores the vehicle's numberplate details allowing the user to pay on exit with a credit card without using a ticket at all. Other systems allow a 'card in – card out' payment whereby the credit card is inserted at entry and exit and payment calculated. This type of system will be put at some risk with the advent of Chip + Pin credit cards because the amount of time that a user will spend in the entry/exit lane while entering their pin number will be significantly increased, reducing the entry/exit capacity of the car park. In places such as airports, where parking charges are high and hence using a pin is likely to be mandatory, this is a significant design issue.

10.7.4   MOBILE PHONE PARKING

Interest in the use of mobile phones as a means of paying for on-street parking is increasing rapidly and there are now a number of systems in use throughout Europe. The principle is simple, the mobile phone is used to communicate the desire to park and payment is collected. There are four basic approaches:

- The mobile phone is used like as a 'remote control' to operate the parking equipment and get it to produce a ticket with payment collected via the phone account. This type of system is in use in Dublin in Eire, where users have to sign up in advance to have their phone identified by the system.
- The phone user opens an account with a bureau and the phone is used to call the bureau and report a parking Act. The nearest pay machine number is given to identify the location and the appropriate transaction is charged to the subscriber.
- An in-car unit is used which uses the mobile phone network to communicate. The user pre-purchases (or more correctly hires) the unit and opens an account. When parking the unit is activated and the charge levied against the appropriate account. This system which was developed in Israel is in use in Holland.
- A system where the user texts their registration mark to a text address, which identifies the parking zone. The message is recorded and the payment collected. The user is sent a text message receipt as proof of payment.

The systems have been in use for a relatively short time and the need for pre-subscription for most of the systems has limited the attractiveness of this form of parking payment. However, a version of the last of the four systems was introduced in Zagreb in Croatia in mid-2001. Called M Parking[14] the system has been expanded to most of the major towns and cities in Croatia and in Zagreb some 52% of parking payments are now made by mobile phone.

Mobile phone parking is being introduced as a complimentary system in location where some other form of payment mechanism is already in place. However, in late 2003, M Parking was introduced in the City of Belgrade as the primary parking system. The city has implemented the mobile phone system as the main means of paying for parking, with a relatively low number of pay and display machines provided as a back stop for those without a mobile phone.

References

1. UK Government (2002) *The Traffic Sign Regulations and General Directions* SI 2002/3113, HMSO, London.
2. Department of Transport (1984) *Road Traffic Regulation Act, 1984*, HMSO, London.
3. Department of Transport (1989) *Local Authorities' Traffic Order (Procedure) (England and Wales) Regulations 1989*, SI 1989/1120, HMSO, London.

4. Department of Transport (1993) *Local Authorities' Traffic Order (Procedure) (England and Wales) (Amendment) Regulations 1993*, Sl 1993/1500, HMSO, London.
5. Institution of Highways and Transportation & Institution of Structural Engineers (1984) *Design Recommendations for Multi-Storey and Underground Car Parks* (2nd edn), Institution of Structural Engineers, London.
6. The Institution of Civil Engineers (2002) *Recommendations for the Inspection, Maintenance and Management of Car Park Structures*, Thomas Telford, London.
7. UK Government (1995) *The Disability Discrimination Act 1995*, HMSO, London.
8. Association of Chief Police Officers (updated) *Secured by Design*, a Scheme for Secure Car Parks, ACPO/AA.
9. British Standards Institution (to be published) *BS 6571 Part 8 Parking Discs*, BSI.
10. British Standards Institution (1989) *BS 6571 Part 1 Clockwork Parking Meters*, BSI.
11. British Standards Institution (1989) *BS 6571 Part 2 Electronic Parking Meters*, BSI.
12. British Standards Institution (1989) *BS 6571 Part 3 Pay and Display Parking Meters*, BSI.
13. British Standards Institution (to be published) *BS 6571 Part 7 Parking Vouchers* BSI, London.
14. British Parking Association (2002) M-Parking, Parking by Mobile Phone in Croatia.

11

Road Safety Engineering

11.1 Factors resulting in accidents

There are three factors that result in accidents:

- road and environment deficiencies;
- road user errors (human factors);
- vehicle defects.

Road and environment deficiencies account on their own for only 2% of all accidents but in combination with road user errors account for slightly less than 20%. Human factors on their own account for 75% of accidents.

Typical road and environment deficiencies are those which provide misleading visual information, or insufficient or unclear information to the road user. Only occasionally accidents are caused solely by bad design.

Human factors include excessive speed for the conditions, failing to give way, improperly overtaking or following too close and general misjudgement by both driver and pedestrian.

11.2 Road accident definition

The two basic types of road accident, which by definition have to involve a vehicle, are:

- personal injury;
- damage only.

A personal injury accident (PIA) is an accident involving an injury. The PIA refers to the accident as the event, and may involve several vehicles and several casualties (persons injured). The accident must occur in the public highway (including footways) and become known to the police within 30 days of its occurrence. The vehicle need not be moving and it need not be in collision with anything.

A casualty is a person killed or injured in an accident. Casualties are subdivided into killed, seriously injured and slightly injured. The definitions for these three subdivisions of severity are:

- *Killed*: A casualty who dies within 30 days of the accident but excluding confirmed suicides.
- *Seriously injured*: An injury for which a person is detained in hospital as an in-patient, or any of the following injuries whether or not they are detained in hospital: fractures,

concussion, internal injuries, crushing, severe cuts and lacerations, severe general shock requiring medical treatment, injuries causing death 30 or more days after the accident.
- *Slightly injured*: An injury of a minor character such as a sprain, bruise or cut which is not judged to be severe, or slight shock requiring roadside attention. This definition includes some injuries not requiring medical treatment.

An injured casualty is recorded as serious or slightly injured by the police on the basis of information available within a short time of the accident. This generally will not include the results of a medical examination, but may be influenced by whether the casualty is hospitalised or not. There is clearly some scope for lack of uniformity of recording.

A fatal PIA is one in which there is at least one killed casualty. A serous PIA is one with no persons killed but at least one seriously injured casualty. A slight PIA is one with no persons killed or seriously injured but at least one slightly injured casualty.

Drivers have legal obligations for reporting accidents but the reporting does not necessarily have to be to the police. The police only record accidents which are reported to them. On average only 70% of accidents involving personal injury are recorded by the police. The recording of PIAs involving only cyclists and no motorised vehicle is about 25%.

Transport Research Laboratory (TRL) has determined that there are, in reality, many damage-only accidents to every accident involving personal injury.

The police are the primary source of information on road accidents. Other sources include:

- motor insurance companies;
- hospital casualty (accident and emergency) units.

11.3   Police recording

The basic accident data itself is collected by the police, being recorded in a notebook using a local standard format. In most cases this takes place as a result of attending a road accident, but there are occasions when the accident will be reported after the event at a police station. An example of a local standard format is shown in the Hampshire Constabulary 'Accident Reporting Systems' as in Figure 11.1.

There are 51 police authorities in Great Britain (GB) and the methods of handling accident data vary considerably between them. Each police force will usually have their own particular form for the accident data, which allows for completion of the STATS19 form (to be described later) and the additional information that is passed to the local authority. The additional information includes contributory factors (e.g. excessive speed) which in the police view have contributed to the accident. In almost all police authorities, the accident data will be entered into a computer database and certain validation checks are carried out. This may be done locally at individual police stations, or there may be a central unit to process the data for the whole police area. In some cases local highway authority staff may be involved.

The issue of the accuracy of the police accident data is one that is often debated between a highway authority and the respective police authority. While it is clearly important for the data to be as accurate as possible, collecting the information is not the first priority when police officers attend the scene of an accident. The extent of reporting of accidents to police stations after the event must also be taken into account, especially as often the police themselves do not go to the site.

Some police authorities record and keep limited information on damage-only accidents which are reported to them, or which are attended. In these cases, there is an additional source of data that may be available to highway authorities to assist in identifying locations for engineering remedial treatment.

 **Hampshire Constabulary**

 T1B   9/93

INJURY ROAD COLLISION REPORTS
Notes for guidance for the completion of road collision reports – T1A
Shaded sections are for computer operator use only

 1   ## ATTENDANT CIRCUMSTANCES

Form Number
If only one form is used, enter 1. For additional forms number consecutively.

Collision Refernece Number
To contain Divisional/Section identification letters and collision reference number. No reference to year is required.

Date
Unused boxes to the left of the day or month are to be entered as zeros; thus 9th May 1992 is coded 09 05 92.

Time
Use the 24 hour clock.

Total Number of Casualties/Vehicles
Enter total number of casualties in collision.
Enter total number of vehicles in collision.

Contributory Factors
Enter up to three factors selected from list overleaf.

Speed Limit
Enter speed limit applicable on the road. Speed limits which are temporarily in force should not be included.

Local Authority
Consult Stats 20, para. 1.10

1st Road Class and Number
Enter the class and number of road on which the collision occurred. If unclassified enter 'U'.

Marker Post Number
Enter kilometres and tenths, using leading zeors as necessary. Tenths should be entered in final box.

2nd Road Class and Number
For collisions occurring at junctions only.

Light Conditions
Codes 1–3 apply to daylight collisions.
Codes 4–7 apply to darkness collisions.

2   ## CASUALTY DETAILS
Each casualty should be numbered consecutively.
(i.e. 1,2,3 on form 1 and 4,5,6 on form 2, etc.)

Vehicle Reference Number
To identify the vehicle occupied by a casualty prior to the collision.

Pedestrian casualty records should quote the vehicle reference number of the vehicle by which the pedestrian is first hit.

Casualty Severity
Fatal injury includes only those cases where death occurs in less than 30 days as a result of the collision. Fatal does not include death from natural causes or suicide.

Serious injury – examples include: fracture, internal injury, severe cuts and lacerations, crushing, concussion, severe general shock requiring hospital treatment, detention in hospital as an in-patient and injuries to casualities who die 30 or more days after the collision.

Slight injury – examples include: sprains, bruises, cuts judged not to be severe, slight shock requiring road side attention.

Car Passenger
In case of drivers, tick '0 – not car passenger.'

DoT Special Projects Local Special Projects
Use only when directed.

School Pupil
Tick box 1 *only* if the casualty is a school pupil on the journey to or from school. All other casualties, including school age children *not* going to or from school, tick box 0.

School pupils are children aged between 5 and 15 years inclusive.

NB
First row or column refers to the first casualty/vehicle on form. Second row or column refers to second casualty/vehicle on form. Third row or column refers to third casualty/vehicle on form.

 3   ## VEHICLE DETAILS

Vehicle Reference Number
Each vehicle is to be numbered consecutively (i.e. 1,2,3 onform 1 and 4,5,6 on form 2, etc.)

Type of Vehicle
The following types of vehicle are defined:
Box 8 – Taxi	–	any licensed hackney vehicle carrying appropriate local authority plates; but does not include private hire vehicles.
Box 11 – Bus or Coach –		any passenger carrying vehicle (PVC) equipped to carry 17 or more persons.
Boxes 12 and 13	–	goods vehicles are classified here by maximum gross weight; either up to and including 3.5 tonnes or over 3.5 tonnes.

Postcode
Give first half of postcode of driver's home address, if known, e.g. if code is S022 5DB enter S022. Part code, e.g. SP is acceptable if remainder is not known. Do not guess if code is known.

Other Vehicle Hit
Enter the vehicle reference number of the first other vehicle (if any) with which the vehicle being coded collided.

DoT Special Projects Local Special Projects
Use only when directed

Overshoot/Restart
Indicate whether vehicle stopped and restarted at junction (restart) or failed to stop at junction (overshoot).

Part(s) Damaged
Up to three codes may be ticked in this section for each vehicle.

Breath Test
In cases where the breathalyser procedure cannot in law be applied, e.g. in the case of non-motor vehicles. tick '0 – not applicable.'

Where a negative breath test has been given T28A *must* also be submitted.

Fig. 11.1 Example of an injury road collision report.

11.4   The National Accident Reporting System and STATS19

The National Accident Reporting System defines and records accidents as only PIAs on the public highway (including footway) in which a vehicle is involved and which became known to the police within 30 days of its occurrence.

The vehicle does not need to be moving and it need not be in collision with anything. A PIA could then include cases of accidents involving people boarding or alighting from vehicles or of injuries when vehicles brake.

The designation of the forms used for the National Accident Reporting System is STATS19. STATS19 has a fixed format and consists of three records which must be completed.

The Accident Record Attendant Circumstances describe the physical nature of the location of the accident. The Vehicle Record contains information on the vehicle involved (but not the registration number) and the driver, with a separate record for each vehicle. The Casualty Record includes more variables dealing with pedestrian casualties than vehicle driver and occupant casualties, to deal with such information as pedestrian location and movement, with a separate record for each casualty involved. These forms are included as Figures 11.2–11.4.

The only personal information shown in STATS19 is age and gender of casualties and drivers, and thus it is not affected by the Data Protection Act. There are linkages between the three records, for example, to identify in which particular vehicle each vehicle casualty was travelling.

The publication *STATS20: Instructions for the Completion of Road Accident Report Form STATS19* is the manual which defines each of the variables in detail, as well as showing how the STATS19 form should be completed.

The STATS19 data is passed by the police through local highway authorities to the Department of Transport (DfT) on a monthly basis, when it is checked and validated, before being added to the national database. Every 3 months, a bulletin is published, showing brief provisional quarterly accident and casualty figures, compared with the same period for the preceding year. An annual DfT report[1] has been published for a number of years, showing comprehensive analyses of the accident variables in terms of accidents and casualties for the latest year, together with the analysis of trends.

11.5   Data transfer to the highway authority

In addition to the fixed format STATS19 data, local highway authorities need to collect other accident information. The exact nature and extent of this additional information will depend upon the arrangements that have been agreed with the respective police authority(s). However, for road safety engineering, it is essential that there is a 'description' and 'location' of the accident in plain language, in addition to the STATS19 information. In many circumstances, the description is vital in piecing together the events leading up to the accident, particularly where there is conflicting evidence from the STATS19 variables.

In rural areas, it is not always possible to determine the location from the grid reference alone and, furthermore, the location allows the original grid co-ordinates to be checked. Other information can be used in some road safety engineering studies, for example the school attended in the case of child pedestrian casualties. In many cases, the police will also include what is called a 'contributory or causation factor', which is a subjective assessment as to the primary cause of an accident, with often similar 'factors' allocated to each casualty and vehicle. The Hampshire Constabulary's Contributory Factors are attached as Figure 11.5; each Constabulary has its own list of factors.

On receipt of the data each month from the police, a number of automatic computer validation checks will be carried out. In many cases, local highway authorities will check the grid co-ordinates manually, with perhaps some other basic manual checking being carried out.

Accident Record Attendent Circumstances

Department of Transport

1.1 Record Type [1] `1 2`
  1 Authorised accident record
  5 Amended accident record

1.2 Police Force [ ] `3 4`

1.3 Accident Ref No. [ ][ ][ ][ ][ ][ ][ ] `5 6 7 8 9 10 11`

1.4 Severity of Accident [ ] `12`
  1 Fatal 2 Serious 3 Slight

1.5 Number of Vechicles [ ][ ][ ] `13 14 15`

1.6 Number of Casualty Records [ ][ ][ ] `16 17 18`

1.7 Date
  Day `19 20` [ ][ ]
  Month `21 22` [ ][ ]
  Year `23 24` [ ][ ]

1.8 Day of Week [ ] `25`
  1 Sunday 2 Monday
  3 Tuesday 4 Wednesday
  5 Thursday 6 Friday
  7 Saturday

1.9 Time 24 hour
  Hrs `26 27` [ ][ ]
  Mins `28 29` [ ][ ]

1.10 Local Authority [ ][ ][ ] `30 31 32`

1.11 Location
  10 digit reference No.
  Easting `33 34 35 36 37` [ ][ ][ ][ ][ ]
  Northing `38 39 40 41 42` [ ][ ][ ][ ][ ]

1.12 1st Road class [ ] `43`
  1 Motorway
  2 A (M)
  3 A
  4 B
  5 C
  6 Unclassified
  7 Local
  8 Authority
  9 Use only

1.13 1st Road Number [ ][ ][ ][ ] `44 45 46 47`

1.14 Carriageway Type or Markings [ ] `48`
  1 Roundabout (on circular highway)
  2 One way street
  3 Dual carriageway – 2 lanes
  4 Dual carriageway – 3 or more lanes
  5 Single carriageway – single track road
  6 Single carriageway – 2 lanes (one each direction)
  7 Single carriageway – 3 lanes (two way capacity)
  8 Single carriageway – 4 or more lanes (two way capacity)
  9 Unknown

1.15 Speed Limit mph `49 50 51` [0][ ][ ]

1.16 Junction Detail `52 53` [0][ ]
  0 Not at or within 20 metres of junction
  1 Roundabout
  2 Mini-roundabout
  3 'T' or staggered junction
  4 'Y' junction
  5 Slip road
  6 Crossroads
  7 Multiple junction
  8 Using private drive or entrance
  9 Other junction

Junction accidents only

1.17 Junction Control [ ] `54`
  1 Authorised person
  2 Automatic traffic signal
  3 Stop sign
  4 Give way sign or markings
  5 Uncontrolled

1.18 2nd Road Class [ ] `55`
  1 Motorway
  2 A (M)
  3 A
  4 B
  5 C
  6 Unclassified
  7 Local
  8 Authority
  9 Use only

1.19 2nd Road Number [ ][ ][ ][ ] `56 57 58 59`

1.20 Pedestrian Crossing Facilities `60 61` [ ][0]
  0 No crossing facilities within 50 metres
  1 Zebra
  2 Zebra crossing controlled by school crossing patrol
  3 Zebra crossing controlled by other authorised person
  4 Pelican
  5 Other light controlled crossing
  6 Other sites controlled by school crossing patrol
  7 Other sites controlled by other authorised person
  8 Central refuge – no other controls
  9 Footbridge or subway

1.21 Light Conditions [ ] `62`
  DAYLIGHT
  1 Street lights 7 metres or more high
  2 Street lights under 7 metres high
  3 No street lighting
  4 Daylight street lighting unknown
  DARKNESS
  5 Street lights 7 metres or more high (lit)
  6 Street lights 7 metres or more high (lit)
  7 No street lighting
  8 Street lights unit
  9 Darkness street lighting unknown

1.22 Weather [ ] `63`
  1 Fine (without high winds)
  2 Raining (without high winds)
  3 Snowing (without high winds)
  4 Fine with high winds
  5 Raining with high winds
  6 Snowing with high winds
  7 Fog (or mist if hazard)
  8 Other
  9 Unknown

1.23 Road Surface Condition [ ] `64`
  1 Dry
  2 Wet/Damp
  3 Snow
  4 Frost/Ice
  5 Flood (surface water over 3cms (1 inch) deep)

1.24 Special Conditions at Site [ ] `65`
  0 None
  1 Automatic Traffic Signal-out
  2 Automatic Traffic Signal partially defective
  3 Permanent road signing defective or obscured
  4 Road works present
  5 Road surface defective

1.25 Carriageway Hazards [ ] `66`
  0 None
  1 Dislodged vehicle load in carriageway
  2 Other object in carriageway
  3 Involvement with previous accident
  4 Dog in carriageway
  5 Other animal in carriageway

1.26 Overtaking Manoeuvre Patterns [ ] `67`
  No longer required by the Department of transport

1.27 DTp Special Projects [ ][ ][ ][ ] `68 69 70 71`

Fig. 11.2 Accident record attendant circumstances.

Vehicle Record

2.1 Record Type
```
  1 2
[ 2]
```
1 New vehicle record
5 Amended vehicle record

2.2 Police Force
```
  3 4
[   ]
```

2.3 Accident Ref No.
```
  5 6 7 8 9 10 11
[              ]
```

2.4 Vehicle Ref No.
```
 12 13 14
[       ]
```

2.5 Type of Vehicle
```
 15 16
[    ]
```
01 Pedal cycle   02 Moped
03 Motor scooter   04 Motor cycle
05 Combination   06 Invalid Tricycle
07 Other three-wheeled car 08 Taxi
09 Car (four wheeled)
10 Minibus/Motor caravan
11 PSV
12 Goods not over 1 1/2 tons UW (1.52 tonnes)
13 Goods over 1 1/2 tons UW (1.52 tonnes)
14 Other motor vehicle
15 Other non motor vehicle

2.6 Towing and Articulation
```
 17
[  ]
```
0 No tow/articulation   1 Articulated vehicle
2 Double/multiple trailer 3 Caravan
4 Single trailer   5 Other tow

2.7 Manoeuvres
```
 18 19
[    ]
```
01 Reversing   02 Parked
03 Waiting to go ahead but held up
04 Stopping   05 Starting
06 U turn
07 Turning left   08 Waiting to turn left
09 Turning right   10 Waiting to turn right
11 Changing lane to left
12 Changing lane to right
13 Overtaking moving vehicle on its offside
14 Overtaking stationary vehicle on its offside
15 Overtaking on nearside
16 Going ahead left hand bend
17 Going ahead right hand bend
18 Going ahead other

2.8 Vehicle Movement Compass Point
From   To
```
 20 21
[    ]
```
1 N   2 NE   3 E
4 SE   5 S   6 SW
7 W   8 NW

or
```
[ 0][ 0]  Parked – not at kerb
[    0]   Parked – at kerb
```

2.9 Vehicle Location at time of Accident
```
 22 23
[    ]
```
01 Leaving the main road
02 Entering the main road
03 On main road
04 On minor road
05 On service road
06 On lay-by or hard shoulder
07 Entering lay-by or hard shoulder
08 Leaving lay-by or hard shoulder
09 On a cycleway
10 Not on carriageway

2.10 Junction Location of Vehicle at First Impact
```
 24
[  ]
```
0 Not at junction (or within 20 metres/22 yards)
1 Vehicle approaching junction/vehicle parked at junction approach
2 Vehicle in middle of junction
3 Vehicle cleared junction/vehicle parked at junction exit
4 Did not impact

2.11 Skidding and overturning
```
 25
[  ]
```
0 No skidding, jackknifing or overturning
1 Skidded   2 Skidded and overturned
3 Jackknifed   4 Jackknifed and overturned
5 Overturned

2.12 Hit Object In Carriageway
```
 26 27
[    ]
```
00 None
01 Previous accident
02 Road works
03 Parked vehicle – lit
04 Parked vehicle – unit   05 Bridge (roof)
06 Bridge (side)   07 Bollard/refuge
08 Open door of vehicle
09 Central island or roundabout
10 Kerb   11 Other object

2.13 Vehicle Leaving Carriageway
```
 28
[  ]
```
0 Did not leave carriageway
1 Left carriageway nearside
2 Left carriageway nearside and rebounded
3 Left carriageway straight ahead at junction
4 Left carriageway offside onto central reservation
5 Left carriageway offside onto central reservation and rebounded
6 Left carriageway offside crossed central reservation
7 Left carriageway offside
8 Left carriageway offside and rebounded

2.14 Hit Object Off Carriageway
```
 29 30
[    ]
```
00 None   01 Road sign/Traffic signal
02 Lamp post
03 Telegraph pole/Electricity pole
04 Tree   05 Bus stop/Bus shelter
06 Central crash barrier
07 Nearside or offside crash barrier
08 Submerged in water (completely)
09 Entered ditch
10 Other permanent object

2.15 Vehicle Prefix/Suffix Letter
```
 31
[  ]
```
Prefix/Suffix letter or one of the following codes:
0 More than twenty years old (at end of year)
1 Unknown/cherished number/not applicable
2 Foreign/diplomatic
3 Military
4 Trade plates

2.16 First Point of Impact
```
 32
[  ]
```
0 Did not impact
1 Front   2 Back
3 Offside   4 Nearside

2.17 Other Vehicle Hit (VEH Ref No.)
```
 33 34 35
[       ]
```

2.18 Part(s) Damaged
```
 36    37    38
[  ]  [  ]  [  ]
```
0 None   1 Front
2 Back   3 Offside
4 Nearside   5 Roof
6 Underside   7 All four sides

2.19 No. of Axles
```
 39
[  ]
```
No longer required by the Department of Transport

2.20 Maximum Permissible Gross Weight
Metric tonnes (Goods vehicle only)
```
 40 41
[    ]
```

2.21 Sex of Driver
```
 42
[  ]
```
1 Male   2 Female
3 Not traced

2.22 Age of Driver
(Years estimated if necessary)
```
 43 44
[    ]
```

2.23 Breath Test
```
 45
[  ]
```
0 Not applicable   1 Positive
2 Negative   3 Not requested
4 Failed to provide
5 Driver not contacted at time

2.24 Hit and Run
```
 46
[  ]
```
0 Other   1 'Hit and run'
2 Non stop vehicle not hit

2.25 DTp Special Projects
```
 47 48 49 50
[          ]
```

Fig. 11.3 Vehicle record.

Fig. 11.4 Casualty record.

3.1 Record Type `[3]` 1 2
1 New casualty record
5 Amended casualty record

3.2 Police Force `[ ][ ]` 3 4

3.3 Accident Ref No. `[ ][ ][ ][ ][ ][ ][ ]` 5 6 7 8 9 10 11

3.4 Vehicle Ref No. `[ ][ ][ ]` 12 13 14

3.5 Casualty Ref No. `[ ][ ][ ]` 15 16 17

3.6 Casualty Class `[ ]` 18
1 Driver or Rider
2 Vehicle or pillion passenger
3 Pedestrian

3.7 Sex of Casualty `[ ]` 19
1 Male
2 Female

3.8 Age of Casualty `[ ][ ]` 21
(Years estimated if necessary)

3.9 Severity of Casualty `[ ]` 22
1 Fatal
2 Serious
3 Slight

3.10 Pedestrian Location `[ ][ ]` 23 24
00 Not pedestrian
01 In carriageway crossing on pedestrian crossing
02 In carriageway crossing within zig-zag lines approach to the crossing
03 In carriageway crossing within zig-zag lines exit the crossing
04 In carriageway crossing elsewhere within 50 metres of pedestrian crossing
05 In carriageway crossing elsewhere
06 On footway or verge
07 On refuge or central island or reservation
08 In centre of carriageway not on refuge or central island
09 In carriageway not crossing
10 Unknown

3.11 Pedestrain Movement `[ ]` 25
0 Not pedestrian
1 Crossing from drivers nearside
2 Crossing from drivers nearside – masked train by parked or stationary vehicle
3 Crossing from drivers offside
4 Crossing from drivers offside – masked by parked or stationary vehicle
5 In carriageway stationary – not crossing (standing or playing)
6 In carriageway stationary – not crossing (standing or playing) – masked by parked or stationary vehicle
7 Walking along in carriageway facing traffic
8 Walking along in carriageway back to traffic
9 Unknown

3.12 Pedestrian Direction `[ ]` 26
Compass point bound
1 N
2 NE
3 E
4 SE
5 S
6 SW
7 W
8 NW
or 0 – pedestrian – standing still

3.13 School Pupil Casualty `[ ]` 27
0 Not a school pupil
1 Pupil on journey to/from school
2 Pupil NOT on journey to/from school

3.14 Seat Belt Usage `[ ]` 28
0 Not car or van
1 Safety belt in use
2 Safety belt fitted – not in use
3 Safety belt not fitted
4 Child safety belt/harness fitted – in use
5 Child safety belt/harness fitted – not in use
6 Child safety belt/harness not fitted
7 Unknown

3.15 Car Passenger `[ ]` 29
0 Not a car passenger
1 Front seat car passenger
2 Rear seat car passenger

3.17 PSV Passenger `[ ]` 30
0 Not a PSV passenger
1 Boarding
2 Alighting
3 Standing passenger
4 Seated passenger

3.17 DTp Special Projects `[ ][ ][ ][ ]` 31 32 33 34

A: Collisions due to actions of drivers/riders

    1 Tired or asleep
    2 Illness
    3 Drunk or drugged
    4 Speed too great for prevailing conditions
    5 Failing to keep to nearside
    6 Overtaking improperly on nearside
    7 Overtaking improperly on offside
    8 Failing to stop at pedestrian crossing
    9 Turning round carelessly
  10 Reversing carelessly
  11 Failing to comply with traffic sign
        (other than double white line or traffic lights)
  12 Failing to comply with double white lines
  13 Starting from nearside carelessly
  14 Starting from offside carelessly
  15 Changing traffic lanes carelessly
  16 Cyclist riding with head down
  17 Cyclists more than two abreast
  18 Turning left carelessly
  19 Turning right carelessly
  20 Opening doors carelessly
  21 Crossing road junction carlessly
  22 Cyclist holding another vehicle
  23 Not in use
  24 Misjudging clearance
  63 Failing to comply with traffic lights
  64 Incorrect use of vehicle lighting

B: Collisions due to actions of pedestrians

  26 Crossing road masked by vehicle
  27 Walking or standing in road
  28 Playing in road
  29 Stepping or running into road carelessly
  30 Physical defects or illness
  31 Drunk or drugged
  32 Holding onto vehicle

C: Vehicle lighting

  33 Dazzle by other vehicle's lights
  34 Inadequate rear lights
  35 Inadequate front lights
  36 Not in use

D: Collisions due to actions of passengers

  37 Carelessly boarding or alighting from bus
  38 Falling inside or from vehicle
  39 Opening door carelessly
  40 Negligence by conductor of bus

E: Collisions due to actions of animals

  41 Dog in carriageway
  42 Other animal in carriageway

F: Collisions due to obstructions

  43 Stationary vehicles dangerously
        placed
  44 Other obstructions

G: Collisions due to defective vehicles

  45 Defective brakes
  46 Defective tyres or wheels
  47 Defective steering
  48 Unattended vehicle running away
  49 Insecure load
  50 Other defects

H: Collisions due to road conditions

  51 Pot hole
  52 Defective manhole cover
  53 Other road surface conditions
  54 Road works in progress
  55 Slippery road surface (not weather)
  65 Flood

I: Collisions due to weather conditions

  56 Fog or mist
  57 Ice, frost or snow
  58 Strong wind
  59 Heavy rain
  60 Glaring sun

J: Collisions due to other factors

  61

K: Collisions due to unknown factors

  62

USE 61 OR 62 ONLY IF ALL OTHER
FACTORS ARE INAPPROPRIATE

Fig. 11.5 Contributory factors.

There are some common accident systems that are used by a number of highway authorities, but in many cases there will be a specific purpose designed system.

The source of accident data for the users of this data, be they local authority officers or consultants, is normally the highway authority. Sometimes the information can only be obtained by applying to the police. In London, the London Research Centre, now part of GLA, holds the data on behalf of TfL and the boroughs. Joint data units carry out a similar function in other metropolitan areas.

11.6   Other data to be collected for safety work

Road safety engineers will normally wish to collect several other items of information. These include:

- up-to-date 1:1250 (or 1:2500) mapping
- classified traffic counts or turning movements
- annual average traffic volumes
- pedestrian flows
- traffic speeds
- population
- street inventory (traffic signals, pedestrian crossings, banned turns, one-way streets, pedestrian guardrail)
- local traffic attractors (e.g. shops, offices, hospitals, schools).

11.7   Presenting accident data for analysis

11.7.1   REGRESSION TO THE MEAN

It is a truism to say that accidents are rare events subject to random variation. Random variation will have a biasing effect if combined with a tendency to select sites for treatment on the basis of accident records in the recent past.

This is because a selection process based on recent past accident records tends to produce sites for treatment which happen to be at a peak in their fluctuation of accident frequency over time. We should expect such sites to experience reductions in accident frequency in subsequent years, regardless of whether remedial measures are implemented.

The effect is commonly known as 'regression to the mean'.

It is desirable therefore to present accident data for analysis over a long period (at least 3 years) and to compare with control sites of a similar nature in the same area. Highway authorities find it difficult to conduct control programmes because of the requirements of the Road Traffic Act.

11.7.2   GRAPHICAL

Presenting accident data on a map background provides an immediate identification of the locations; that is, junctions and road sections, where accidents have occurred. Digitised mapping is now commonly used.

It is useful to record the accidents separately as slight, serious or fatal so that the graphical presentation can also provide a measure of severity by location.

11.7.3   STICK DIAGRAMS

When a particular location is to be studied in more detail, it is useful to present the information as a stick diagram as in Figure 11.6. The vehicle manoeuvre for each accident often has to be added manually using symbols which vary by authority.

MVA REFERENCE NO:	2	8	47	111	7	74			
DAY	THURSDAY	TUESDAY	THURSDAY	FRIDAY	SUNDAY	SATURDAY			
DATE	11/08/83	13/09/88	21/12/89	08/01/93	04/09/88	29/12/90			
TIME	1250	1420	1235	1745	1630	1740			
DARK/L				DARK					
WET/D			WET	WET		WET			
SEVERITY	PED/SLIGHT	PED/SLIGHT	PED/SLIGHT	PED/SERIOUS	VEH/SLIGHT	VEH/SLIGHT			
CONFLICT		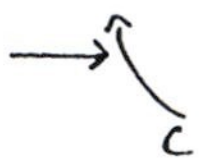							
AGE OF CASUALTY	13								
CONTRIB F	D	D	OC		U	I			
EASTING	517950	517940	517950	517950	517950	517950			
NORTHING	104390	104400	104400	104400	105200	104390			

Fig. 11.6 Accident analysis used by West Sussex County Council Urban Safety Management.

Year ending	No. of accidents		
30.4.83	3	FAT	0
		SER	2
		SLT	1
30.4.84	6	FAT	0
		SER	3
		SLT	3
30.4.85	3	FAT	0
		SER	2
		SLT	1
Total	12	FAT	0
		SER	7
		SLT	5

Accident severity by year

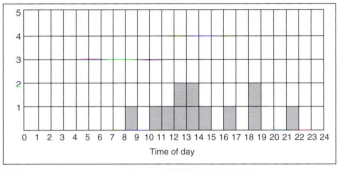

Vehicle category	No.	%	Reg. %
Car/Taxi	13	50	63.6
Goods	11	42.3	8.8
PSV	0	0	3.1
Motor cycle	2	7.7	13.7
Pedal cycle	0	0	6.3
Other	0	0	4.5
Total	26	100%	100%

Vehicle category

Casualty class	No.	%	Reg. %
Driver			54.5
	20	100	
Passenger			26.7
Pedestrian			18.8
Total	20	100%	100%

Surface	No.	%	Reg. %
Dry	10	83.3	62
Wet	2	16.7	33.9
Snow	–	–	4.1

Road surface condition

	No.	%	Reg. %
Skid	1	8.3	27
No skid	11	91.7	73

Skid/no skid

Age	No.	%	Reg. %
0–15	0	0	15.5
16–19	3	15	22.4
20–29	10	50	24.3
30–59	6	30	27.3
60+	1	5	10.5

Casualty age

	No.	%	Reg. %
Daylight	12	100	67
Darkness		0	33

Daylight/darkness

Fig. 11.7 Accident histograms.

11.7.4   *HISTOGRAMS (CONTINGENCY TABLES)*

Histograms are useful when studying a location to identify special accident features, for example:

- vehicle category
- time of day
- month of year
- day of week
- dry, wet or icy road surface
- daylight or darkness (lit or unlit)
- age of casualty
- type of casualty.

The histogram can include national averages for comparison purposes as in Figure 11.7.

11.7.5   *TYPE OF ACCIDENT*

A series of graphical presentations can provide information in more detail for the same area. The detail could include vehicle type and vulnerable road user (pedestrian, cyclist) by age or a particular type of accident (e.g. overtaking).

11.8   Summarising data

11.8.1   *THE 1987 ROAD SAFETY TARGETS*

Highway authorities will want to summarise accident data for the whole of their area to compare with national or local road safety targets. The national road safety target set in 1987 was:

> A one-third reduction in all road casualties by the year 2000 with respect to the average level of casualties for the years 1981–1985.

Achievement against this target is shown in Table 11.1. While there was more than a one-third reduction in killed and seriously injured casualties, the overall target was not met. This is primarily due to the growth in slightly injured casualties.

Road accident statistics by region and county are available in 'Road Accident Statistics in English Regions'.[2] Table 11.2 provides some examples of reduction in casualties between the 1981–1985 baseline and 1995.

The 1997 Casualties recorded in Greater London is shown in Table 11.3. This highlights that pedestrian casualties are some 19.8% of the total number of road traffic casualties but pedestrian fatalities are 57.6% of the total number of road traffic fatalities. Pedestrians involved in a road accident are much more likely to be killed than any other traveller.

Table 11.1 Casualties in GB[1]

	1981/1985 baseline	1999	Actual change	Change (%)
Killed	5598	3423	−2175	−39
Seriously injured	74 533	39 122	−35 411	−48
Slightly injured	241 787	277 765	+35 978	+15
All casualties	321 918	320 310	−1608	0

Table 11.2 Selected regional data – reduction in casualties 1981/1985–1995

	Killed and seriously injured	All casualties
Greater London	−20.9	−16.9
Kent	−41.3	−11.9
Hampshire	−49.9	−3.3
Oxfordshire	−63.0	−10.0
South Yorkshire	−43.5	0.2
Warwickshire	−37.2	8.6
West Midlands	−40.4	−2.6
West Sussex	−34.0	−4.8
England	−38.2	−2.5
Wales	−44.7	3.9
Scotland	−40.0	−18.2
GB	−38.7	−3.5
Northern Ireland	−29.0	42.9
UK	−38.4	−2.4

Table 11.3 1997 casualties – mode of travel by severity (Greater London)

Mode of travel	Severity of casualty			Total	Total (%)
	Fatal	Serious	Slight		
Pedestrian	159	1975	6969	9103	19.8
Pedal cyclist	12	562	3805	4379	9.5
Powered two wheeler (TWMV)	32	958	5486	6476	14.1
Car	61	2754	19 362	22 177	48.2
Taxi	1	33	363	397	0.9
Bus or coach	2	250	1916	2168	4.7
Goods vehicle	5	131	959	1095	2.4
Other vehicle	4	51	165	220	0.5
Total	276	6714	39 025	46 015	100
Total (%)	0.6	14.6	84.8	100	

Greater London has performed better than the rest of GB in that casualty levels in 1997 were 15% lower than the average for the 1981/1985 baseline. The Greater London data for 1997 are shown in Table 11.4 disaggregated by age group and sex.

The percentage change in accidents by type of road user in GB[4] is shown in Tables 11.5 and 11.6 with the current rates in GB for killed and seriously injured casualties by road user shown in Table 11.7. While the figures show considerable reduction for pedestrians and cyclists, those campaigning for vulnerable road users suspect that roads have not become safer for those two groups of travellers but busier and more dangerous, and therefore have put the would-be pedestrians and cyclists off using them.

Table 11.4 1997 casualties – mode of travel by age group and sex (Greater London)

Mode of travel	Age group					Sex		Total
	0–15	16–24	25–59	60+	Unknown	Male	Female	
Pedestrian	2561	1414	3321	1276	531	5206	3897	9103
Pedal cyclist	775	785	2453	138	228	3498	881	4379
Powered two wheeler	35	1394	4750	84	213	5879	597	6476
Car	1440	5280	12855	1469	1133	11039	11138	22177
Taxi	8	29	256	59	45	256	141	397
Bus or coach	213	147	749	790	269	696	1472	2168
Goods vehicles	41	202	770	40	42	942	153	1095
Other vehicle	14	24	127	28	27	151	69	220
Total	5087	9275	25281	3884	2488	27667	18348	46015
Total (%)	11.1	20.2	54.9	8.4	5.4	60.1	39.9	100

Table 11.5 All casualties by road user (GB)

	1981/1985 baseline	1999	Actual change	Change (%)
Pedestrians	61742	41682	−20060	−32
Pedal cyclists	28391	22124	−6267	−22
TWMV	65193	26192	−39001	−60
Car users	143944	205735	+61791	+42
Bus and coach	10182	10252	+70	0
Goods vehicles	11196	10608	−588	−5
All road users	321918	320310	−1608	0

TWMV: two-wheeled motor vehicles.

Table 11.6 Killed and seriously injured casualties by road user (GB)

	1989	1999	Change (%)
Pedestrians	17296	9678	−44
Pedal cyclists	5100	3122	−39
TWMV	12488	6908	−45
Car users	29684	20368	−31
Bus and coach	835	611	−27
Goods vehicles	2673	1407	−47
All road users	68531	42545	−38

TWMV: two-wheeled motor vehicles.

Table 11.7 Passenger casualty rates (KSI) by mode of travel

Mode	Rate per 100 million		
	Passenger kilometres	Passenger journeys	Passenger hours
Car	4.5	55	190
Van	2.4	30	75
TWMV	150.0	1600	4500
Pedal cycle	85.0	230	1200
Foot	55.0	55	225
Bus or coach	1.6	13	40

TWMV: two-wheeled motor vehicles.

Some of the reasons for the national success in reducing the numbers of fatal and seriously injured casualties are:

- safer cars;
- greater compliance with seat-belt regulations;
- safer roads – improved design and safety audit and use of road safety engineering measures;
- reduction in drink/driving;
- changing attitudes with travellers more aware of the dangers of accidents.

11.8.2   2000 ROAD SAFETY TARGETS

DETR's March 2000 document 'Tomorrow's Roads – Safer for Everyone' sets out new 10-year targets and launches a new national road safety strategy. The targets are that by 2010, in comparison with the average for 1994–1998, the UK achieves:

- 40% reduction in the number of people killed or seriously injured (KSI) in road accidents;
- 50% reduction in the number of children KSI;
- 10% reduction in the slight casualty rate, expressed as the number of people slightly injured per 100 million vehicle kilometres.

There are 10 main themes:

- safer for children
- safer drivers (training and testing)
- safer drivers (drink, drugs and drowsiness)
- safer infrastructure
- safer speeds
- safer vehicles
- safer motorcycling
- safer pedestrians, cyclists and horse riders
- better enforcement
- promoting safer road use.

Despite having overall one of the safest road accident records in the world, the UK record on child pedestrian accidents is poor as shown in Table 11.8. This is the reason for specifically including a target for reducing child accidents.

Table 11.8 Child pedestrian fatality rates per annum per 100 000 children

Country	Children 0–14 pedestrian fatality rates
GB	1.21
Austria	0.79
Belgium	0.94
Denmark	0.85
Finland	0.94
France	0.91
Germany	0.64
Ireland	1.31
Italy	0.49
Netherlands	0.66
Norway	0.81
Spain	0.94
Sweden	0.54
Switzerland	0.96

Table 11.9 Summary of changes in casualties for London Target Categories by Year 2000

	Target change by 2010 (%)	Casualties 1994–1998 average	Casualties 2000	Change (%) by 2000 compared with 1994–1998 average
KSI Casualties				
Total	−40	6684	6117	−8
Pedestrians	−40	2137	1870	−12
Pedal cyclists	−40	567	422	−26
Powered two wheelers	−40	933	1195	+28
Children	−50	935	728	−22
Slight casualties				
Total	−10	38 997	39 770	+2

The incidence of child pedestrian casualties is much higher in deprived areas. Children in the most deprived 10% of wards in England are more than three times as likely to be pedestrian casualties than their counterparts in the least deprived 10% of wards, according to a study undertaken by the Institute for Public Policy and Research.

The Mayor's Transport Strategy set additional targets for London intended to promote and increase walking and cycling, and also to recognise the recent (last 4 years) increase in the use of powered two wheelers in London.

Table 11.9 shows a summary of the changes in casualties in London compared with the 1994–1998 average and the target reduction to be achieved by the year 2010.

11.9 Data presentation and ranking

11.9.1 THE FOUR APPROACHES

The basis of the UK structured system[3] for accident investigation and prevention is the use of four main approaches:

- *Single sites*: Treatment of specific sites or short lengths of road where clusters of accidents have occurred.
- *Mass action plans*: Application of a known remedy to locations having common accident factors.
- *Route action plans*: Application of remedies to a road having an above average accident rate for that type of road and traffic volume.
- *Area action plans*: Remedial measures over an area with an above average accident rate for that type/size of area or population.[3]

The method of presentation of results, ranking and numerical/statistical analysis depends on the approach.

11.9.2 ACCIDENT TOTALS AND RATES

Sites, for example junctions, can be listed in order of numbers of accidents. Road sections can be listed in rate order as numbers of accidents per 100 million vehicle kilometres. Areas can be listed in rate order as numbers of accidents per square kilometre or population. Listing accidents in rate order instead of number order can identify sites of high risk for lower numbers of users.

11.9.3 POTENTIAL ACCIDENT REDUCTION

Another way of ranking for treatment is to present the accident data for each site/road section/area in terms of potential accident reduction (PAR). This can be used in combination with totals and rates for choosing top sites for analysis.

11.9.4 TYPE OF ACCIDENT OR USER

For mass action plans the engineer will be concentrating on ranking for a particular type of accident or road user. The type of accident could be skidding, dark (night-time) hours or red-light running for action plans associated with skid-resistance surfacing, improved lighting or introduction of red-light cameras. Red-light running is the disobeyance by drivers of the red aspect at traffic signals.

The type of road user could be pedestrian or cyclist with action plans for introducing pedestrian and cyclist safety improvements.

11.9.5 ACCIDENTS PER SQUARE KILOMETRE OR POPULATION

For area action plans the engineer will be ranking areas by identifying accidents per square kilometre or resident population statistic. This could be used for ranking areas of accidents of a similar nature (e.g. involving vulnerable road users) for traffic calming treatment.

11.9.6 COMPARISON WITH AVERAGES

National road accident statistics, obtained from the national database, are available in various government documents. An example extraction[4] is shown as Table 11.10.

Regional and county averages are published by the highway authorities, London Research Centre and joint data units. These averages are useful for providing a comparison against accident rates on roads in the area being analysed.

Table 11.10 PIA rates by road type and year per 100 million vehicle kilometres

Road type	Year (19..)																
	83	84	85	86	87	88	89	90	91	92	93	94	95	96	97	98	99
Motorways	12	12	12	13	11	11	11	11	10	11	11	11	10	11	11	11	11
Built-up roads (speed limit of 40 mph or less)																	
A roads	130	133	128	124	115	112	108	106	97	95	95	94	93	91	94	93	91
Other roads	136	138	129	121	111	110	99	100	95	90	86	85	82	83	82	82	79
Non-built-up roads (speed limit exceeding 40 mph but excluding motorways)																	
A roads	38	37	37	37	34	35	34	33	29	29	28	28	27	27	27	27	25
Other roads	50	49	47	47	42	44	44	44	42	40	44	45	44	45	44	44	42
All roads	83	82	78	75	67	67	63	62	57	56	55	55	53	53	53	52	50

11.10   Statistical analysis

11.10.1   SIGNIFICANCE LEVELS

A significance level for some accident statistic of $x\%$ implies that there is only a $x\%$ chance of that statistic occurring due to random fluctuations. We can refer to a significance level of $x\%$ or confidence level of $(100 - x)\%$ in the following terms:

Significance level (%)	Confidence level (%)	Significance
1	99	Highly significant
5	95	Significant
10	90	Fairly significant
>20	<80	Not significant

11.10.2   POISSON TEST

This is a useful test for calculating the probability of a particular number of accidents occurring at a location in a given year when the long-term average of numbers of accidents is known. For example, if the average annual number of accidents at a location is two and then in one year five accidents occur, this can be shown using Poisson tables to be significant. This is because the probability of five or more accidents occurring due to random fluctuations is only 5% and so there is a significant (95%) chance that a real increase in accidents has occurred.

11.10.3   CHI-SQUARED TEST

This is a test for determining whether the number of accidents of a certain type at a particular site are significantly different to the number at similar (control) sites. It is also used for determining whether the number of accidents at a site after remedial measures have been carried out has changed significantly with reference to similar control sites over the same period.

Using the following table:

	Numbers of accidents	
	Before measures	After measures
Treated site	*b*	*a*
Control site	*B*	*A*

If χ is >4 then the treated site is significantly different at the 5% significance level.

$$\chi = \frac{(bA - aB)^2(a + b + A + B)}{(b + a)(B + A)(b + B)(a + A)}$$

11.11 Problem identification

11.11.1 LOOKING FOR COMMON PATTERNS

The road safety engineer will make use of the accident data presented to him to identify common patterns. Typical patterns would be:

- above average night-time accidents associated with poor or non-existent lighting;
- skidding accidents associated with poor surfacing or drainage;
- pedestrian accidents associated with inadequate crossings or excessive vehicle speed.

Problem identification is not a mechanical exercise and so has to be learnt from practical experience. It is often necessary to use several different methods of presentation and to refer to the traffic data and street inventory.

The engineer would normally supplement this analysis of common patterns to identify problems by one or more of the following:

- on-site observations
- conflict studies
- location sampling.

On-site observations
These involve driving, walking and observing the location over an extended period of time, and possibly at different times of the day and in different weather and lighting conditions.

Photographs taken at driver eye height or pedestrian eye height provide a record of visual perception.

The site visit can, among other matters, also identify:

- poor sight lines (and intervisibility between traffic flows) caused perhaps by street furniture;
- confusing situation through poor road markings or inadequate maintenance of signs;
- forward visibility impaired by foliage or parked vehicles.

Conflict studies
This involves observations of vehicle movements at specific locations in order to assess the frequency and type of 'near accident' situations.

Location sampling

Location sampling is useful in developing mass action plans. It involves the grouping of accident data for sites with similar physical or accident features to provide sufficient data for the assessment of contributory factors (e.g. excessive speed).

11.11.2   RECENT RESEARCH

Some recent research undertaken by the Automobile Association Foundation for Road Safety Research with Cambridgeshire County Council has identified some common aspects of accidents on rural roads that relate to road environment and engineering issues:

1. The incidence of accidents is not a simple function of traffic volume or of flow to capacity ratio.
2. A carriageway widening right-turn facility (ghost island) at T-junctions reduces the number of stacking-type accidents in the approach lane remote to the minor road though they can cause an increase in accidents for vehicles turning right out of the minor road. An untested remedy could be to change the road markings to provide a substandard 1-metre right-turn facility while retaining the extra carriageway width.
3. Accidents at T-junctions with busy private access are a concern but the highway authority has few powers to treat them. Selective carriageway widening at these T-junctions is likely to have the greatest reduction effect.
4. Accidents at T-junctions involving a high proportion of dry-weather skidding imply excessive speed.
5. Skidding accidents on wet roads are highest on roads with a poor level of skidding resistance.
6. The probability of an overtaking-type accident increases if the forward visibility is considerably greater than the design standard. These accidents are more likely to occur if sight distance is substandard (<215 metres) or very good (>580 metres).

TRL has undertaken considerable research on road safety and some guidance on aspects of the safety of link and junction arrangements can be obtained from that research. The London Accident Analysis Unit have also produced reports on topics covering accidents specific to the elderly, children, parked vehicles, minibuses, taxis, pedestrians, public service vehicles, pedal cyclists, signal-controlled junctions, younger drivers and vehicle speed.

11.12   Institution of Highways and Transportation accident investigation procedures

The attached Figure 11.8 shows the components of the Institution of Highways and Transportation (IHT) procedures:[5]

- data provision (data collection);
- problem site and situation identification (presenting accident data for analysis, ranking and numerical and statistical analysis);
- diagnosis of problem (problem identification);
- selection of site for treatment.

This fourth procedure area, selection of site for treatment, involves identification of possible remedies and choice of remedy with assessment of accident benefit.

Further guidance is available from Royal Society for Prevention of Accidents (RoSPA)[6] and the European Transport Safety Council.[7]

Procedure areas   Tasks   Ranking stages

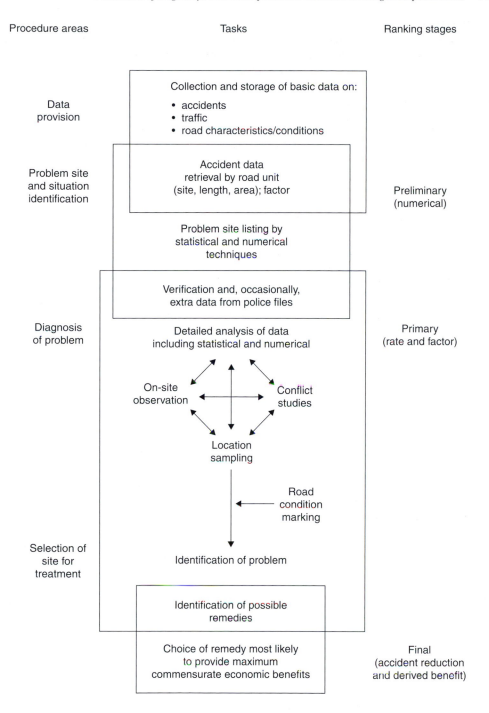

Fig. 11.8 Outline of procedures for identification, diagnosis and selection of sites.

11.13   Designing road safety engineering measures

Once ranking of sites for treatment has been undertaken, then engineering measures to remedy the accident problem will be designed and accident savings assessed. Engineering measures will be prepared not only to remedy accident problems but also to improve perception of safety; for example, to overcome the deterrent to walking, particularly among the young, the less mobile and the elderly caused by heavily trafficked roads with inadequate crossings for pedestrians.

The measures to be considered depend on the approach adopted, for example:

- single site treatment
- mass action plans
- route action plans
- area action plans.

Typical examples of measures are:

- Single site treatment
  - improved signing
  - carriageway markings
  - road surface treatment
  - lighting improvements
  - alterations to alignment, kerbs and islands
  - introduction of signal control or mini-roundabouts.
- Mass action plan
  - lighting improvements
  - anti-skid surfacing
  - speed enforcement cameras
  - red-light running cameras
  - pelicans, zebras and pedestrian phases at signals.
- Route action plan
  - carriageway widening at junctions
  - speed limits
  - speed control measures
  - side road closures or left-in, left-out only
  - cycle routes.
- Area action plan
  - vertical deflection – humps and tables
  - horizontal deflection – chicanes and narrowings
  - mini-roundabouts
  - road entry treatment
  - road closures and banned turns
  - 20 mph zones.

Figures 11.9 and 11.10 show before and after views of road layouts improved as accident prevention measures.

11.14   Accident savings

The procedure for assessing accident savings is pragmatic; there is no textbook definitive formula for assessment. General guidelines are available from the latest research and studies undertaken either by or on behalf of IHT, TRL and local highway authorities. TRL have identified that

(a)

(b)

Fig. 11.9 Accident prevention scheme at a rural crossroads
(courtesy Hampshire County Council): (a) before and (b) after.

(a)

(b)

Fig. 11.10 Accident prevention scheme in a village centre
(courtesy Hampshire County Council): (a) before and (b) after.

achievement of a 1 mph reduction in average vehicle speeds can reduce accidents by as much as 5%. IHT have suggested the following guidelines for percentage accident reductions:

Single site action	33%
Route action	15%
Redistribution and general treatment	10%
20 mph zones	50–75%

Reductions due to redistribution and general treatment require further explanation. Accident reduction can be achieved by measures designed to redistribute traffic within an area by actively discouraging the use of certain routes, which may not have particular accident problems but will benefit from a reduction in vehicle flow.

A reduction can also be achieved by general treatment of different parts of the study area network to adapt the behaviour of traffic to match the primary function of the streets through which it is passing. This is found to be particularly important where traffic redistribution is not entirely feasible but where accident risk is not great enough to justify a site-specific measure. These measures are intended to complement the other measures of the scheme and often have their greatest value in satisfying local residents' concerns. Redistribution and general treatment, when read together in an overall package of schemes, can expect to produce a saving of 10%. This is in addition to the savings expected from single site action.

Where feasible, accident reduction should be assessed by the likely effects of specific treatments. For example, if a treatment measure is designed to reduce right-turn accidents at a junction where these are particularly high, then the saving can be taken as the reduction to the normal rate for that junction type. Guidance on accident rates at different junction types is available from TRL.

11.14.1   THE COST OF ACCIDENTS

Department for Transport valuation of the costs of PIAs highlights the value of problem identification. These costs at June 1999 prices are (from Highways Economics Note No. 1) as follows:

Injury	Cost (£)
Fatal accidents	1 253 140
Serious accidents	146 890
Slight accidents	14 540
All PIAs	49 800
Damage-only accidents	1300
Average per PIA allowing for damage-only accidents	69 270

These values reflect 'standard gamble' surveys of people's 'willingness to pay' which gives a higher value than that based on medical expenses and lost economic output.

The accident benefit of a measure is the number of accidents saved multiplied by the appropriate cost per accident.

The value of an accident remedial scheme is usually represented by its first year rate of return, which should normally be at least 15% for consideration for funding:

$$\text{First year rate of return} = \frac{\text{accident benefit for one year}}{\text{cost of implementation}}$$

11.15   Road safety plans

Virtually all local highway authorities produce annual road safety plans and guidance on their production is provided by the Local Authority Association.[8]

These plans set out the authorities' strategy for road safety and plans for safety measures. These will cover road safety engineering measures proposed but will also cover other issues, such as:

- accident data summary
- road user education
- publicity
- enforcement
- safety targets
- co-ordination
- encouragement of safety awareness
- monitoring.

11.16   Road safety audits

Safety audits[9,10] are primarily intended to ensure that new road schemes, improvements to highways and traffic management measures are designed, and implemented to operate as safely as possible.

Road safety audits are mandatory for all Highways Agency, Scottish Executive, Welsh Assembly and Department for Regional Development Northern Ireland trunk road schemes, and are good practice for measures on all other roads.

The audits identify potential safety hazards typically under different grades of severity, for example 'problem' or 'warning'. The auditor or audit team report to the client project manager who will, when necessary, then instruct the scheme design team to respond with alternative designs.

Audits are undertaken at each of four stages for trunk road schemes:

- completion of preliminary design before publication of draft orders;
- completion of detailed design before invitation to tender;
- completion of construction;
- monitoring.

The scope of the audit becomes progressively more detailed between stages one, two and three. Items covered include:

- departures from design standards
- alignment
- junction layout
- provision for pedestrians and cyclists
- signs and lighting
- buildability
- road markings
- safety fences.

11.17   Effect of speed

11.17.1   BREAKING THE SPEED LIMIT

The 1999 Speed Survey (refer to Transport Statistics Bulletin, Vehicle Speeds in Great Britain, 1999) showed that on urban roads:

- with a 30 mph speed limit 67% of cars exceeded the speed, 31% travelling faster than 35 mph. On 40 mph roads 26% of cars exceeded the limit, with 8% exceeding 45 mph;

- motorcycles were the vehicles most likely to be speeding on 40 mph roads, with 38% exceeding the speed limit and 22% doing so by >5 mph. On 30 mph roads 38% exceeded 35 mph;
- with 30 mph speed limit, 45% of articulated heavy goods vehicles (HGVs) exceeded the speed limit, 13% by >5 mph.

Results for rural roads show that:

- Cars were the vehicles most likely to be speeding on motorways with 56% of those surveyed exceeding 70 mph and 19% travelling in excess of 80 mph.
- On non-urban dual carriageways 53% of cars exceeded 70 mph and 14% over 80 mph and on non-urban single carriageway roads 10% of cars exceeded the 60 mph limit, 2% travelling over 70 mph.
- On dual carriageways 90% of articulated HGVs surveyed exceeded their 50 mph limit. On single carriageways levels of speeding were generally considerably lower although HGVs are still high with 76% of articulated HGVs exceeding their 40 mph limit.

11.17.2   EFFECT OF SPEED ON ROAD SAFETY[11]

The effect of excessive speed on road safety has been clearly identified. For some time, it has been recognised that a reduction of 1 mph in average road speeds can be expected to cause a reduction of 5% in accident frequency.

11.18   Monitoring performance of remedial measures

11.18.1   CONTROL SITES

Any monitoring of the performance of remedial measures is likely to involve the use of control sites to remove the effect of changes in numbers of accidents in the area due to other events (e.g. area accident trends overtime or changes in traffic flow).

Control sites should be of a similar nature to the site being monitored; for example, control sites could be all other signalised junctions in the same town as the signalised junction site subject to remedial measures.

11.18.2   EVALUATING EFFECTIVENESS

Some of the criteria that can be used for evaluating effectiveness of remedial measures are:

- accident numbers
- traffic speeds
- traffic flows
- public perceptions.

11.18.3   STATISTICAL/NUMERICAL TESTS

Statistical or numerical tests of effectiveness in reducing accident numbers include:

- chi-squared test
- standard error
- *K* test.

The chi-squared test has already been described. Standard error methods identify the size of the effectiveness.

The *K* test is a simple numerical test using data from control sites:

$$K = \frac{a/b}{A/B}$$

where

> *b*: before accidents at site,
> *a*: after accidents at site,
> *B*: before accidents at control sites,
> *A*: after accidents at control sites.

If $K < 1$ then the site has had a decrease in accidents relative to the control sites.

11.18.4   MOLASSES

The MOLASSES (Monitoring Of Local Authority Safety SchemES) project was initiated by the County Surveyors' Society's 'Accident Reduction Working Group' in 1991, in an attempt to encourage more monitoring of safety engineering work undertaken by highway authorities.

The objectives of the project were:

- to develop a central computer database for building up information about the effectiveness of safety engineering schemes implemented by Local Authorities within the UK;
- to compile data received on schemes into the database;
- to provide information for County Surveyors Society (CSS) reports and for individual authorities, if requested;
- to provide software for data transfer and record keeping.

The database contains detailed information on schemes and treatments and can provide both information on how well different treatments worked and what operational difficulties were encountered, so that experience can be shared and, hopefully, best designs can be achieved. All Highway Authorities were asked to contribute to the database and the information will become increasingly useful as the coverage of schemes becomes more comprehensive.

The database is not intended as a vehicle for statistical research, such a system would be complicated and expensive to maintain. However, it does provide a tool for collating the results obtained for a particular type of treatment in a reasonably systematic way, and provided caution is exercised when interpreting the results, it can provide a useful source of information on experience with different measures.

In total, 34 authorities have contributed to the MOLASSES database. The number of schemes in the database now stands at 2300 and, on average, about 60 schemes a month have been entered into the database.

Schemes entered into the MOLASSES database are classified based on the type of treatment installed at a site. To make the provision of information on treatments simpler and quicker, the MOLASSES input form provides a set of treatment codes, which describe the treatment types employed at specific sites.

To demonstrate the potential of the MOLASSES database for analysis of the effectiveness of various treatments, 20 of the most commonly occurring treatments were examined in terms of the effect on 'all' accidents, and also the effect on 'target' accidents where they had been identified separately.

Table 11.11 lists the treatment codes examined, in rank order of the percentage reduction of 'all' accidents. The number of schemes representing several of the treatments is relatively small as only single treatment sites, or sites where the specific treatment is the dominant feature, have been included. In addition, only sites with 2–5 years comparable accident data in the 'before' and 'after' installation periods have been included.

Accident data have been grouped from all sites with the same treatment codes, with accident reductions calculated from this grouped data in the 'before' and 'after' periods.

Overall, the effect on accidents of the schemes shown in Table 11.11 has been very good, with a reduction in all accidents of 53% and a reduction in target accidents of 73%. The effect of individual schemes varied considerably, with a few cases actually seeing an increase in accidents. The overall averages given must thus be treated with some caution. Individual design and site-specific effects are obviously also important.

None of these results take into account any 'regression to mean' effect, which could be present if the sites had been selected for treatment on the basis of 'before' accident figures which were higher than the underlying mean. It is likely that some 'regression to mean' effect will have occurred, thus reducing the savings quoted, but there is not sufficient information available on

Table 11.11 'Percentage' and 'total' accident reductions for the most common treatment types

Treatment	Number of sites	Percentage accident reduction		Total reduction			
				'All' accidents		'Target' accidents	
		All	Target	Before	After	Before	After
3.3 – Priority junction visibility	11	73	73	62	17	22	6
2.2 – New mini-roundabout	6	71	81	45	13	27	5
5.0 – Pedestrian facility anti-skid	7	71	33	69	20	15	10
3.1 – Priority junction geometry	14	69	76	157	49	37	9
7.31, 8.31 – Link/route anti-skid	13	68	53	148	47	32	15
5.5 – Pedestrian refuges	17	68	71	192	61	79	23
3.0 – Priority junction RH turn lanes	22	68	51	210	67	49	24
3.4 – Priority junction signing	7	68	84	28	9	25	4
7.22, 8.22 – Link/route signing	12	65	63	72	25	30	11
4.5 – Bend kerbing	21	61	88	140	54	49	6
4.2 – Bend visibility	15	58	66	74	31	53	18
7.21, 8.21 – Link/route re-surfacing	16	57	72	211	91	123	34
4.4 – Bend signing	14	57	70	68	29	50	15
5.0 – Pedestrian barriers	6	54	100	42	19	20	0
7.27 – Link lighting	6	53	56	66	31	27	12
1.2.1 – Signal junction new ped phase	5	53	100	71	33	34	0
5.2 – New pelican crossing	25	48	58	157	81	77	32
1.1 – New signal junction	11	38	100	124	77	24	0
7.20 – Link carriageway markings	20	29	65	215	152	69	24
7.23, 8.23 – Link/route signs and marking	21	27	94	449	326	97	6
Total	269	−53	−73	2600	1232	939	254

how the sites were selected to make a precise estimate of how large such a 'regression to mean' effect might be.

References

1. Department of Transport (annual) *Road Accidents Great Britain – The Casualty Report*, HMSO, London.
2. Department of Transport (annual) *Road Accident Statistics in English Regions*, HMSO, London.
3. Institution of Highways and Transportation (1990) *Guidelines for Urban Safety Management*, IHT, London.
4. Department of Transport (annual) *Transport Statistics Great Britain*, HMSO, London.
5. Institution of Highways and Transportation (1990) *Highway Safety Guidelines – Accident Reduction and Prevention*, IHT, London.
6. Royal Society for Prevention of Accidents (1992) *Road Safety Engineering Manual*, RoSPA, London.
7. European Transport Safety Council (1995) *Reducing Traffic Injuries Resulting from Excess and Inappropriate Speed*, EC, Brussels.
8. Local Authority Association (1989) *Road Safety Code of Good Practice*, LAA.
9. Department of Transport (2003) Design Manual for Roads and Bridges HD 19/03, *Road Safety Audits*, HMSO, London.
10. Institution of Highways and Transportation (1996) *Guidelines on Road Safety Audit*, IHT, London.
11. DETR (2000) *New Directions in Speed Management, A Review of Policy*, March, London.

12
Traffic Calming

12.1 Objectives

Traffic calming has two main objectives: the reduction in numbers of personal injury accidents and improvement in the local environment for people living, working or visiting the area.

Traditional traffic management uses physical measures and legislation to coerce, and mould driver behaviour to coax higher capacities out of the highway network, with improved levels of safety. Traffic calming now uses an expanded repertoire of measures and techniques to change driver's perception of an area. Many streets portray the impression that they are vehicular traffic routes that have some other uses of lesser importance, such as shopping streets or for residential access. Traffic calming can alter the balance and impress upon the driver that the street is primarily for shopping or residential use and that vehicular traffic is of secondary importance.

Regardless of the prime cause of accidents, it has long been recognised that there is a direct correlation between accident severity and vehicle speed. Excessive speed for the prevailing road conditions can in itself be the prime cause of some accidents. Speeding traffic can cause severance effects between two parts of a community due to the difficulties experienced when pedestrians attempt to cross the road. High speed vehicles produce high noise levels and consequently degrade the environment. The Transport Research Laboratory (TRL) has identified that a 3–7% reduction in accidents can be expected for every 1 kilometre/hour reduction in vehicle speeds in urban areas.

Increasingly traffic calming is being seen as part of the urban design toolkit. This means that traffic calming designs must be consistent with improvements to the urban streetscape. The traffic engineer must modify his designs to be sensitive to streetscape imperatives which will vary by the type of area. For example, the measures appropriate in an industrial area may well not be visually acceptable in a residential area or shopping street.

12.2 Background

Almost from the dawn of the motor age, transport planners and policy makers have assigned a hierarchy to the road network with inter-urban trunk roads, primary distributors, district distributors, local distributors and access roads. In urban areas, increasing vehicular traffic levels and congestion has eroded the differences between the road types. Longer distance traffic has diverted to 'rat-runs' through local areas and traffic flows have grown on secondary roads to levels formerly associated with primary routes.

Highway engineers have always tried to design roads to appropriate standards for their position in the hierarchy and road safety was always uppermost among their objectives. In the new towns this approach can be seen clearly. In residential areas, narrow roads with tightly curving alignments lead to wider carriageways with larger radii, until the national motorway or trunk road network is reached. Most of the basic network of roads in the older cities, towns and villages in Britain was established long before motorised transport. In the Victorian era there was a massive expansion in the sizes of settlements and wide, straight grid-like street patterns were constructed.

Diversion of traffic to parallel routes adjacent to congested routes was successfully tackled by traffic engineers in the 1970s and 1980s, using a wide range of techniques including road closures, as described in earlier chapters.

In Europe, particularly Holland, Germany and Denmark, where problems were similar to the UK, a different approach was adopted. Many of the UK techniques were used but the basic approach was a re-design of the roads in built-up areas but generally with an emphasis on residential areas. The re-design of streets and areas in Europe has not only reduced speeds and accidents by making the driver feel that he is intruding into an essentially pedestrian environment, but has softened and enhanced the aesthetics of the treated area.

In Britain traffic calming, with a few exceptions, has continued as the imposition of traffic engineering measures to reduce speeds and accidents. The concept of re-design has not often been achieved, partly because of the legislative framework in the UK and lack of financial resources but also due to some erstwhile highway and traffic engineers who have been slow to adopt radical new ideas.

Most 'traffic calming' schemes in the UK have involved a form of speed-reducing road hump. Speed humps are, undoubtedly, cheap and effective in reducing vehicle speeds but it is debatable whether they have a calming effect on drivers. The demand for traffic calming schemes continues unabated as does the search for inexpensive measures that are acceptable to drivers and effective in reducing accidents and speeds.

12.3   Site selection and ranking

Limited financial resources means that there will be competition for schemes within a local authority area. There is a danger that councillors and officers will be pressured by those who 'shout loudest' to implement schemes in those areas first. Prioritising of schemes, based upon objective criteria, will not only help to relieve these pressures but should result in resources being applied where most benefit can be achieved.

There are two approaches to producing a ranking for traffic calming, the single site or road and the area-wide approach. In urban areas it is likely that treatment of a single road will have a 'knock on' effect upon parallel routes and other adjacent streets. In small rural towns and villages where there are no nearby parallel routes, the single site can be examined and assessed in isolation.

For assessment purposes, the single site can be used successfully provided there is a clear understanding of the diversionary effect and the area affected. Preliminary cost estimates which might be used in the ranking list would need to cover the treatment of the diversionary network. If a 'single site' is chosen for implementation it would 'trigger' treatment of the wider network if this is necessary.

The area-wide approach would apply to an identifiable area or cell within the road network. A significant problem with the area-wide approach is that local accident or environmental 'hot spots' can be so diluted by a safe, tranquil hinterland that they are ignored. To identify 'hot spots' a more traditional AIP approach might be needed to highlight sites for treatment. The single-site approach is less likely to miss isolated sites.

Any ranking system must have a common basis for assessment. Single-site assessment can be based upon various rates per kilometre of road. This may, also, be a sound basis for area assessments and will tend to minimise the effects of the size and density of the road network in the area.

The prime objectives for traffic calming schemes vary from each local authority area to another. However, it is probable that limited financial resources will dictate that accident reduction is the most important, but not exclusive, objective. Table 12.1 shows some priority factors that can be

Table 12.1 Priority factors for traffic calming schemes (London Borough of Ealing)

Criterion	Range	Priority factor
A Vehicle Speed mph	over 45	12
(85th percentile speed)[a]	41–45	10
	36–40	8
	31–35	6
	26–30	4
	20–25	2
	under 20	0
B Vehicle flow	Over 1000	10
(vehicles/hour)	900–1000	9
(average for peak hours)	801–900	8
	701–800	7
	601–700	6
	501–600	5
	401–500	4
	301–400	3
	201–300	2
	101–200	1
	under 100	0
C Cyclists (average per hour over four highest hours in any day)	per 10	1
D Pedestrians (pedestrians/km/highest hour over four highest hours in any day) Crossing road	per 100	3
E Number of frontage residents/km	per 100	1
F Accident level (personal injury accidents/km/year averaged over a 3 year period)	per accident	5
	under 1	0
G Special features in area[b]	School entrance	6
	Hospital, Nursing home	3
	Elderly persons home	3
	Bus route	3
	Local shops, Post office, etc.	3
	Doctor surgeries	3
	Community centres	3
	Luncheon clubs	3
	Voluntary play groups	3
	Nurseries	

[a] 85th percentile speed refers to the speed at or below which 85% of traffic is travelling; this is a commonly accepted value for speed assessment studies.
[b] 'Special features' means features which are likely to create particular vehicle, and more particularly pedestrian, movements on the road network.

used to determine a priority ranking for single-site traffic calming schemes. This is similar to the list used by the London Borough of Ealing. It is possible that two parallel routes, that might be mutually affected by calming measures in the other, would appear on the same list. The route with the highest priority factor would trigger measures on both routes (and the intervening/hinterland areas).

The area-wide approach uses a similar method to establish a hierarchy. Both approaches could include more subjective, qualitative criteria particularly where there is less pressure on resources.

The site selection process not only provides essential information to the programming of detailed study and scheme implementation. It can be used, after a settling down period, to provide a measure of the effectiveness of the implemented scheme. A new priority level can be calculated after a year (or preferably 3 years). Hopefully, the scheme, if fully implemented, will have fallen from the priority list and will only require vigilance from the traffic engineer to ensure that the implemented measures do not degrade through unsympathetic or poor maintenance.

12.4   Consultation

The questions here are when? where? who? and how to consult? There is little point trying to impose schemes upon people. It is probable that the ranking procedure will have identified areas with problems which are well known to occupants. As usual there is more than one way to address this problem:

- Ask local people, by means of a questionnaire, their views on various aspects of traffic in their neighbourhood and, possibly, canvass their solutions.
- Present a set of affordable potential options to the local people and ask their views on the options, traffic problems in general and alternative solutions.

The first method ensures that local people feel involved but can receive criticism for not explaining what could be done.

The second method can receive the criticism that local people are being presented with a *fait accompli* and are not being allowed to participate sufficiently.

When a number of affordable options have been designed it is essential to obtain a consensus view from local people. Supporting information in the form of accident plots, speed surveys, traffic flows, photographs of trouble spots, scheme options and diagrams and photographs of features similar to those proposed should be provided. Local people are often interested in the possible after effects and costs and any example sites within reasonable travelling distance from their homes. It should be made crystal clear to consultees that it is an exercise in public participation and not just consultation, and that proposals can and will be adjusted where necessary or even dropped altogether. A carefully constructed questionnaire will help the identification of a preferred option. Consultations must include the emergency services and public transport operators.

12.5   Traffic calming techniques

Traffic calming techniques can be broken down into eight broad categories:

- legislation and enforcement
- surface treatment and signing
- vertical deflection
- horizontal deflection
- gateways and entry treatment

- 20 mph zones
- home zones
- lorry control schemes.

12.5.1   LEGISLATION AND ENFORCEMENT

These methods consist of restrictions on movement and parking including speed restriction, banned turns and one-way streets. In every case the effectiveness of each measure will depend upon the levels of enforcement that are achievable. Traffic orders made under the Road Traffic Regulation Act by the local highway authority are needed to enable enforcement to be carried out.

Enforcement of speed restrictions is aided by speed cameras but, as drivers become more familiar with them, effects are likely to deteriorate to their immediate vicinity. Camera support masts and loop detector installations cost between £5000 to £7000 per site plus the cost of the camera (£15 000–£20 000) and processing of film and prosecutions. It is common for one camera to be moved between up to eight prepared camera sites as most police forces maintain a smaller number of cameras than sites. Speeding motorists will, increasingly, become aware that the probability of being caught is less.

12.5.2   SURFACE TREATMENT AND SIGNING

Surface treatment can consist of simple coloured or textured lengths of carriageway, such as those used for gateway treatments, through to whole blockwork paved streets. Surface treatments can be very effective but generally require reinforcement by more positive measures.

Signing is an essential to give authority to traffic orders but has been used with surface treatments to produce village/town gateways which increases drivers awareness when entering speed limited areas. Signing alone cannot produce all of the calming effects required in an area and its use is governed by the Traffic Sign Regulation and General Directions (TSRGD) 2002.[1] The TSRGD is comprehensive but does not allow for innovative, one-off designs to be used.

12.5.3   VERTICAL DEFLECTION

Most vertical deflections consist of some form of speed-reducing road humps.[2] Road humps were first introduced experimentally in the 1980s and produced immediate speed reductions. The now superseded Highways (Road Humps) Regulations 1990 allowed their use throughout the country at suitable sites and with fairly strict siting criteria. The humps were round or flat topped and a series of humps had to be preceded by a speed-reducing feature (e.g. a 'give-way' line, a sharp bend or roundabout). Many authorities side stepped the speed hump regulations, which were considered too restrictive, and introduced speed tables and platforms where the carriageway is lifted up to footway level, usually in a contrasting colour or textured material such as concrete blocks. The 1990 Regulations brought virtually all vertical deflections under its control. Vertical deflections are now controlled in The Highways (Road Humps) Regulations 1996.[3]

Most bus operators were unhappy with speed humps but accepted long speed tables with ramp gradients shallower than 1:15 (6.7%) or speed cushions.[4,5] Speed cushions are similar to speed humps but are narrower so that wide vehicles can traverse them with little interference. They are usually constructed in pairs but can be used singly or three abreast (Figure 12.1).

Other vertical deflections are: thermoplastic road humps[6] which can be between 900 mm and 1500 mm in length and up to 50 mm high and round topped; and rumble devices[7] which can consist of transverse strips of thermoplastic up to 13 mm high, blockwork or coarse chippings set in epoxy resin (Figure 12.2).

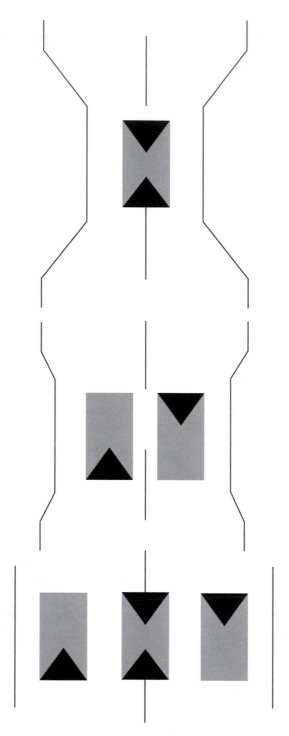

Fig. 12.1 Speed cushion layouts.

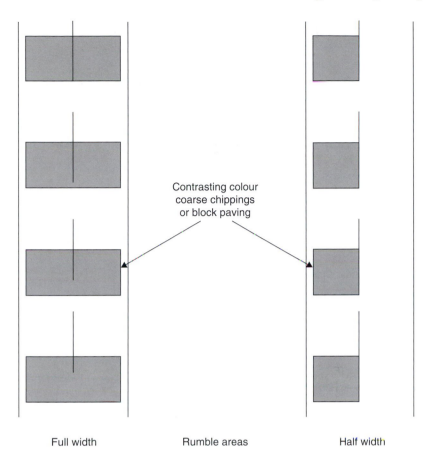

Contrasting colour
coarse chippings
or block paving

Full width   Rumble areas   Half width

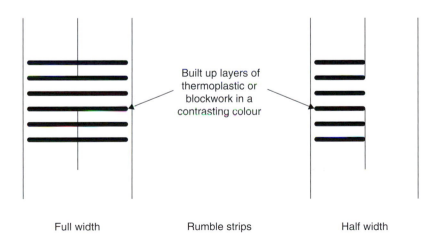

Built up layers of
thermoplastic or
blockwork in a
contrasting colour

Full width   Rumble strips   Half width

Fig. 12.2 Rumble devices.

12.5.4   *HORIZONTAL DEFLECTIONS*

Chicanes or pinch points can effectively reduce vehicle speeds but designs that allow the passage of large vehicles[8] often do not slow light vehicles sufficiently. Careful positioning of traffic islands on the approaches to chicanes can improve their performance but should be used with great care when there is heavy parking pressure. Inconsiderately parked cars can obstruct the passage of buses and other large vehicles. The traffic islands can provide a refuge for pedestrians where parking is restricted and sight lines are unobstructed. Carefully designed parking schemes can be used to produce horizontal deflection of the carriageway using road markings. Adequate enforcement is essential if the beneficial effects are not to be lost.

Priority 'give-way' designs, where one lane on a two-way carriageway is eliminated and vehicles entering an area are forced to yield to exiting vehicles is very effective. Priority 'give-ways' have been used for village 'gateways' and for speed-reducing features at the start of traffic calming schemes (Figure 12.3).

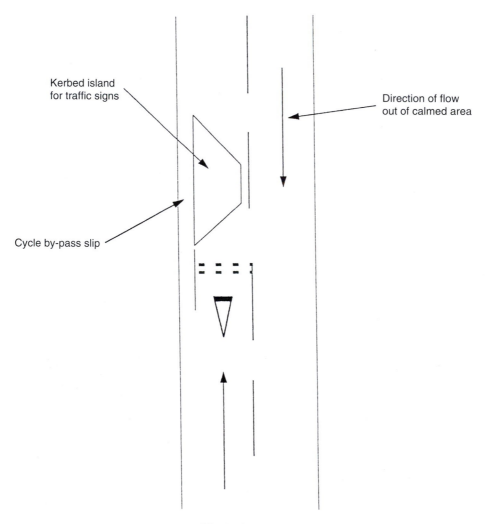

Fig. 12.3 Priority give-way.

12.5.5   GATEWAYS AND ENTRY TREATMENT

Carriageway narrowing combined with suitable signing, street furniture, gateposts, fencing, surface treatment or speed tables can be used to change drivers' perceptions when entering an area (Figure 12.4).[9]

12.5.6   20 MPH ZONES

The lowest national speed limit in urban areas is 30 mph but this might be higher than is appropriate in certain areas. The Department of Transport now allows 20 mph zones[10] to be implemented, generally in residential areas, where measures have been installed to both discourage through traffic and to control the 85th percentile speeds of vehicles entering and using the area to 20 mph. Zone signs must be erected at every entrance to the zone and they must be supported by a traffic order[11] (See Figure 12.5).

12.5.7   HOME ZONES

Home zones[15] are areas (typically residential streets or living areas) where different users share the roadspace. They are allowed in England and Wales under the provisions of the Transport Act 2000. The traffic speeds in a home zone are typically well below 20 mph and allow for pedestrians and cyclists to use the same space as that used by cars and delivery vehicles.

The main benefit is to an improved street environment and social perception though the reduction in speed can bring highway safety benefits with it. Community involvement is required at an early stage in the design of the home zone often using 'planning for real' consultation/involvement techniques.

Home zones are popular with residents but comparatively expensive; often creating a home zone in an area can cost 10 times that of a conventional traffic calming solution. There is then a challenge to justify the additional expense in giving wider benefits to the community than can be realised from improvements in road safety provided by a traffic calming solution.

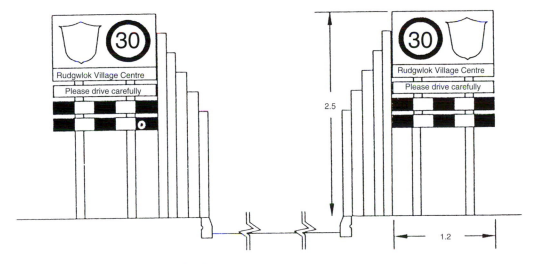

Fig. 12.4 Typical gateway treatment.

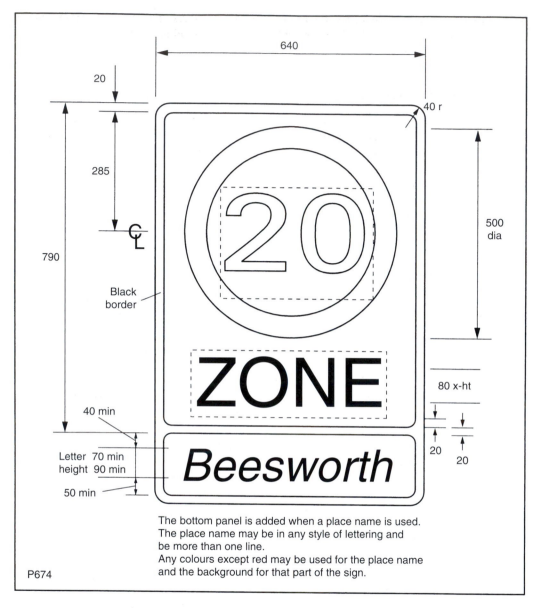

The bottom panel is added when a place name is used.
The place name may be in any style of lettering and
be more than one line.
Any colours except red may be used for the place name
and the background for that part of the sign.

P674

Fig. 12.5 Signing for a 20 mph zone.

12.5.8   *LORRY CONTROL SCHEMES*

Lorry control schemes vary from bans on through traffic using a single street as a 'rat-run' to area-wide control schemes. There are many of the former but only a few area-wide schemes. Two prominent area-wide schemes are the Windsor Cordon, administered by Windsor and Maidenhead Council, and the London Lorry Ban administered by the Association of London Government. The London ban operates throughout the Greater London area and restricts lorries over 18 tonnes maximum gross weight from using all roads in London, except the former trunk road network, overnight, Saturday afternoons and all Sundays. Exemption permits can be obtained if the driver ensures that he does not leave the trunk road network until as near as practicable to the planned stopping place.

12.6   Achievable speed reductions

Vertical defections can, depending on their spacing and severity, be expected to reduce mean speeds to between 15–25 mph. Locally, at the ramp, speeds will be around 15–20 mph. Gentle chicanes and central refuges will reduce mean speeds to below 30 mph and severe chicanes to around 20 mph.

The VISP study (TRL, 1994, paragraph 4.2[12,13]) found that the introduction of a gateway without any speed-reducing features within the village, reduced 85th percentile speeds by between 0 and 7 mph at the gateway but that, generally, speed reductions were not maintained through the village if no other speed-reducing measures were introduced.

12.7   Estimate of accident reductions and benefits

The potential for accident reduction through urban safety management schemes as described in the Institution of Highways and Transportation (IHT) Guidelines (IHT, 1990, Paragraph 5.2[14]) is made up of three parts.

The first part is a reduction in accidents achieved by traditional site-specific schemes as part of a whole area-wide package. Research has indicated that these small scale local safety measures can reduce personal injury accidents by between 15% and 80%, with an overall average of 33%.

The second part is an accident reduction achieved by measures designed to redistribute traffic within an area by actively discouraging the use of certain routes, which may not have particular accident problems but will benefit from a reduction in vehicle flow.

The third part is a reduction achieved by general treatment of different parts of the study area network to adapt the behaviour of traffic to match the primary function of the streets through which it is passing. This is found to be particularly important where traffic redistribution is not entirely feasible but where accident risk is not great enough to justify a site-specific measure. These measures are intended to complement the other measures of the scheme and often have their greatest value in satisfying local residents' concerns.

The second and third parts, when read together in an overall package of schemes, can expect to produce a saving of 10%. This is in addition to the savings expected from the first part.

From these guidelines and the historic accident data, an estimate of the expected accident reductions can be made. If accident savings are translated into monetary values an estimate of economic benefits can be equated with the schemes' capital cost to estimate a first year rate of return (FYRR) on the investment. This method ignores environmental benefits, which are difficult to quantify in monetary terms, and maintenance costs.

12.8   Urban design

Recent years have seen an emphasis on an holistic approach to designing streets. Colin Davis set the debate in motion with his views on how to improve the high street.[16] Some of his key messages were the need to specify quality pavements, and to reduce the street furniture clutter created by signs and guard rail. His message that traffic engineers should consider these issues in detailed design is now accepted.

Since then we have had several publications taking the debate further forward and setting standards for urban design.[17–20] The commission for architecture and the built environment (CABE) and the IHT have taken this forward to create a series of training courses for urban design for traffic engineers and transport planners.

The starting point is that well-designed streets are a fundamental right of everyone and that the potential benefits of better urban design are enormous. With 80% of our public realm being public highway, creating better streets and movement spaces is a high priority.

Traffic calming is an essential element in good design but calming measures need to be provided as part of a design that empowers the local community rather than leaving it subservient to traffic by designing for people and not just for vehicles. Good urban design creates spaces in our urban areas that encourage people to stay and enjoy.

Residential road networks allow permeability by pedestrians and cyclists and the exploitation of the perimeter block form of development with properties facing onto the street. Individual roads/streets are traffic calmed, that is slow speeds encouraged, by providing access roads and shared surfaces.

Access roads should be designed by geometry and layout for a speed of around 20 mph and can be part of a bus route in recognition that buses too need good permeability through a residential area and should not be confined to the external road network.

Shared surfaces, typically leading off access roads, will be designed with appropriate surfacing and street features for a vehicle speed of around 10 mph, so that pedestrians and vehicles can coexist. Shared surface will not have separate footways.

References

1. UK Government (2002) *Traffic Signs Regulations and General Directions 2002*, HMSO, London.
2. Department of Transport (1990) *Speed Control Road Humps*, Traffic Advisory Leaflet 2/90, HMSO, London.
3. Department of Transport (1996) *Highways (Road Humps) Regulations 1996*, Traffic Advisory Leaflet 7/96, HMSO, London.
4. Department of Transport (1998) *Speed Cushions*, Traffic Advisory Leaflet 4/94, HMSO, London and *Speed Cushion Schemes*, Traffic Advisory Leaflet 1/98, DETR .
5. London Transport Buses (1996) *Traffic Calming and Buses*, London Transport Buses, London.
6. Department of Transport (1994) *Thumps Thermoplastic Road Humps*, Traffic Advisory Leaflet 7/94, HMSO, London.
7. Department of Transport (1993) *Rumble Devices*, Traffic Advisory Leaflet 11/93, HMSO, London.
8. Department of Transport (1993) *Overrun Areas*, Traffic Advisory Leaflet 12/93, HMSO, London.
9. Department of Transport (1993) *Gateways*, Traffic Advisory Leaflet 13/93, HMSO, London.
10. Department of Transport (1991) 20 mph *Speed Limit Zones*, Traffic Advisory Leaflet 7/91, HMSO, London and 20 mph *Speed Limits and Zones*, TA9/99, DETR.
11. Department of Transport (1993) 20 mph *Speed Limit Zone Signs*, Traffic Advisory Leaflet 2/93, HMSO, London.
12. Department of Transport (1994) *Village Speed Control Working Group – Final Report*, HMSO, London and *VISP A Summary*, Traffic Advisory Leaflet 1/94, HMSO, London.
13. Department of Transport (2000) *Traffic Calming in Villages on Major Roads*, Traffic Advisory Leaflet 1/00, DETR.

14. Institution of Highways and Transportation (1990) *Guidelines for Urban Safety Management*, IHT London.
15. DTLR (2001) *Home Zones – Planning and Design*, Traffic Advisory Leaflet 10/01.
16. *Improving Design in the High Street*, Colin. J. Davis, Royal Fine Art Commission, London, 1997.
17. Places, streets and movement – a companion guide to design Bulletin 32 (1998) *Residential Roads and Footpaths*, DETR, London.
18. *Movement Access, Streets and Spaces*, Hampshire County Council, Winchester, 2001.
19. *The 2002 Designing Streets for People Report*, UDAL, 2002.
20. *Urban Design Compendium*, English Partnerships, London, 2000.

13
Public Transport Priority

13.1 Design objectives

Public transport priority has to be seen in the context of an overall urban transport strategy with objectives which include not only improved bus (or tram) operation and restraint of car-borne commuting but also an enhanced environment for residents, workers and visitors. Measures proposed must serve all these objectives and yet also be demonstrably cost-effective and enforceable.

Typical design objectives for public transport priority measures include:

- to improve the conditions and reliability of bus operations through the introduction of appropriate bus priority measures;
- to alter the traffic balance in favour of buses at those locations where this can be properly justified;
- to improve conditions for bus passengers at stops and interchanges;
- to improve road safety generally and, in particular, for pedestrians, cyclists and people with disabilities;
- to review, where appropriate, hours of operation of waiting and loading restrictions;
- to establish and implement the co-ordinated and coherent application of waiting, parking and loading enforcement regimes on bus route corridors;
- to improve conditions for all road users and frontagers on bus route corridors.

Achieving these objectives often involves compromises between improving bus operation and the needs of local businesses and residents for reasonable access and of pedestrians and cyclists for safe and convenient movement.

Bus priority measures should be seen as part of the tool kit that will enable the realisation of the transport strategy. The impact of these measures on bus operation can be powerful, yet that impact should not be exaggerated. On their own, bus priority measures are unlikely to cause the major shift in travel from car to bus that is often needed to improve the urban environment. Yet, combined with other measures, bus priority can contribute to a strategy of improving the urban environment and road safety and minimising the need for car travel. Typical other measures include:

- a restrictive city centre parking policy for commuters;
- improved bus services including park and ride;
- improved bus information for passengers;
- more road space provided for pedestrians and cyclists;

- traffic calming measures in residential cells;
- compatible policies for controlling new development in line with planning policy guidance as set out by the DoE.

13.2   The reliable track

The key to improved bus operation is the provision of a reliable track or right-of-way. In urban areas in the UK this track is normally shared with all other traffic and, as a result, buses experience the same congestion. This can be avoided by providing a physically segregated track, such as in busways or bus-only streets, or by reserving sections of otherwise shared track for use by buses. Issues such as whether guided busways should be provided are secondary to the provision of this track.

If the track is so designed that buses can avoid congestion, then bus timings and timetables can be guaranteed with more certainty and passengers provided with a service that can be relied upon. It is the regularity and reliability of the service that is even more crucial to passengers than over-all time savings. Passengers want to know when the bus will arrive and how long it will take to reach its destination. In many congested urban areas, these two items are extremely variable.

Regularity is the variability in arrivals of buses at stops; that is the variability in bus headway. This directly affects the time passengers spend waiting at stops. Reliability is the variability in bus journey time for a given average journey; this directly affects the amount of travel time a passenger must allow to be confident of reaching the destination by a required arrival time.

Figure 13.1 shows an example, provided by a major UK bus operator, of the regularity in actual bus arrival times on a bus route compared with the timetabled arrival times. This shows a typical bus route with a frequency of five buses per hour, that is a scheduled headway of 12 minutes. The theoretical 'perfect' average waiting time for a bus should be half the headway (i.e. 6 minutes).In practice, because of poor regularity, the average waiting time was recorded as 9 minutes. When considering journeys on this bus route, regular daily users would face a wait in excess of 20 minutes at least on 1 day per month.

13.3   Bus priority measures

Typical bus priority measures fall into four main categories:

- bus lanes and busways
- traffic and parking management measures
- traffic signal control
- bus stop improvements.

These are considered separately, but in practice the design for a bus route corridor will draw on measures from all these categories.

13.4   Bus lanes and busways

13.4.1   WITH-FLOW BUS LANES

With-flow bus lanes are relatively commonplace. They enable buses to avoid queues on congested sections of road by providing a lane marked and signed clearly and implemented under a traffic regulation order prohibiting use by general traffic. Normally cyclists and taxis are allowed with buses in the lane. Sometimes commuter coaches are allowed and very occasionally, in special circumstances, goods vehicles will also be allowed (Plate 13.1).

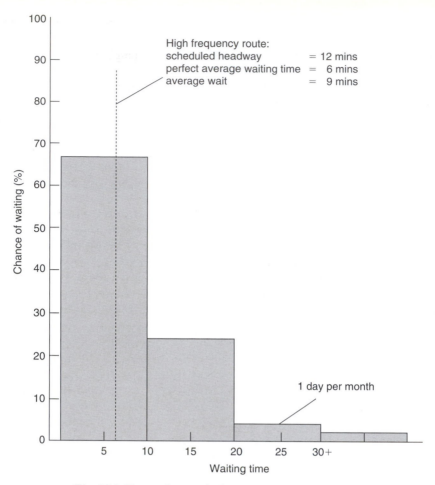

High frequency route:
scheduled headway = 12 mins
perfect average waiting time = 6 mins
average wait = 9 mins

1 day per month

Chance of waiting (%)

Waiting time

Fig. 13.1 Economic appraisal – example of bus regularity.

Plate 13.1 Double decker using a bus lane.

Guidance on road markings and signing can be found in 'Keeping Buses Moving'[1] and 'Traffic Signs Regulations and General Directions'.[6]

Road sections are typically congested for only short periods each day and there is often a need for use of the kerbside for access to properties. For this reason, kerbside bus lanes are often restricted to operate for only a few hours each day (e.g. 7–10 a.m.). Bus lanes to assist right turns would normally operate 24 hours.

The set back, length and set forward of the bus lane are important design considerations; these are illustrated in Figure 13.2. The set back defines the end of the lane and the distance back from the stop line at the next junction. For a roundabout the set back can be short, perhaps three vehicles length, without affecting capacity. For a signalised junction, the set back length in metres is normally at least twice the green time for the approach in seconds to ensure that the approach capacity is not reduced. When capacity is not an issue or when buses share a lane at the stop line with left-turn traffic, then the set back can be shorter or can continue up to the stop line.

The length of the bus lane should be sufficient to ensure that buses avoid being trapped in a queue. If the queue with a bus lane stretches back to a previous signalised junction, then the start of the bus lane should be set forward from that junction a sufficient distance in metres, typically three times the green time in seconds for a two-lane exit, to ensure that exit capacity is not reduced.

The width should be at least 3.0 metres. Usually cyclists are allowed to share the use of the bus lane. However, between about 3.8 and 4.2 metres wide bus drivers might be tempted to overtake a slow moving cyclist; this 'grey area' should be avoided. At 4.3 metres or above an overtaking manoeuvre by a bus is relatively safe, also cyclists can overtake a stopped bus safely within this width. Where buses need to overtake one another, for example at a bus stop, then the width should be increased to 5.5 metres.

Bus lanes can only be provided where the road is sufficiently wide. In a typical urban situation the road should be at least 10.0 metres wide so that a goods vehicle or bus travelling in the opposite direction to the bus lane can safely overtake a stationary bus or loading vehicle. At this restricted width the carriageway should be divided into a 3.0-metre bus lane, a 3.0-metre with-flow general traffic lane and 4.0 metres lane for the opposing flow. Where there is insufficient width for a bus lane but it is important to give priority to buses consideration should be given to using a 'virtual bus lane' as discussed in Section 13.6.4.

13.4.2   CONTRAFLOW BUS LANES

Contraflow bus lanes enable buses to avoid circuitous routes, for example in a gyratory system, by permitting two-way movement for buses only over a road section as illustrated in Figure 13.3. The main disadvantage of a continuous contraflow bus lane is that it prevents kerbside access by vehicles, such as goods vehicles, that are not permitted to use it. A common method of overcoming this disadvantage, particularly when the contraflow bus lane would be long, is to use a point closure to all vehicles except buses and then allow other vehicles access from side roads along its length.

13.4.3   BUSWAYS AND BUS-ONLY STREETS

Busways and bus-only streets provide a dedicated track for use by buses. Busways are either separate roadways or are part of a roadway but segregated from it, normally by kerbs. One-way busways are typically 6.0 metres wide to allow for overtaking broken-down vehicles, and two-way busways are typically 7.3 metres wide.

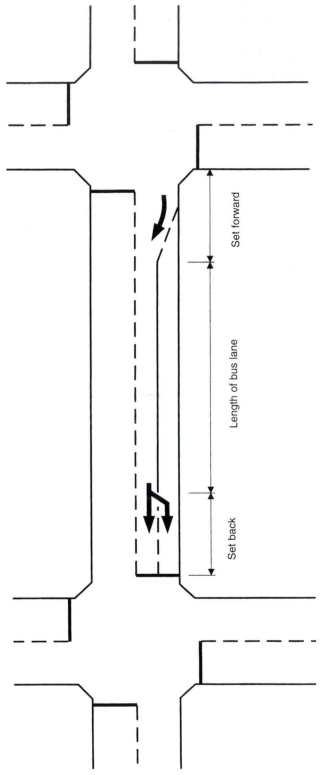

Fig. 13.2 With-flow bus lane.

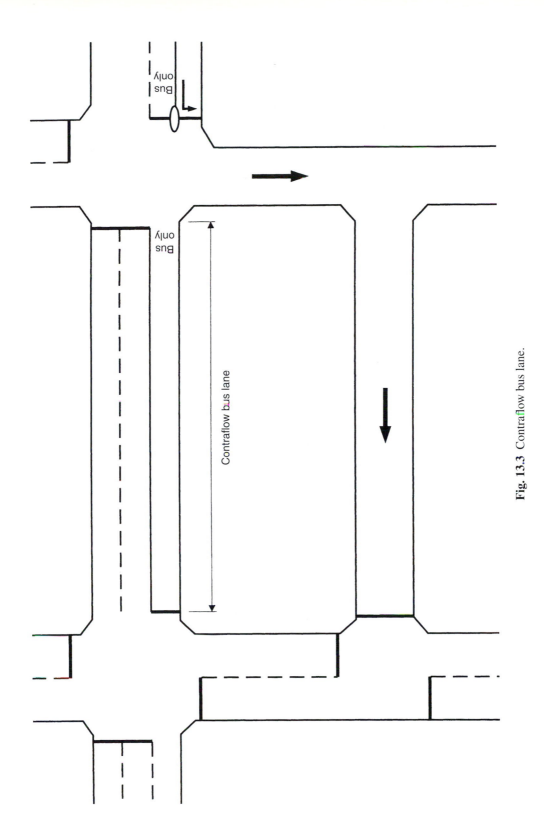

Fig. 13.3 Contraflow bus lane.

Plate 13.2 Guided bus in Leeds.

Guided busways are an attractive option when space is limited, because the width can be reduced as buses are guided either mechanically by horizontally mounted wheels running along kerbs or electronically (Plate 13.2).

Bus-only streets are typically found in town centres; often there will be a speed limit of 10 or 20 mph to ensure that buses travel at speeds compatible with pedestrians. Normally goods vehicles will be permitted access at some period during the day to load/unload at properties. It has been found that some pedestrianised or bus-only town centres are deserted in the evening and cars are now allowed to enter at this time to address this problem.

13.5　Traffic and parking management measures

13.5.1　BUS GATES

'Bus gate' is a generic term describing all forms of control, which allow buses free movement but prevent movement by other vehicles.

They can literally be gates or rising bollards which open/close or lower/raise as a bus approaches and then leaves the gate. Bus gates can also be provided by traffic management measures as now described.

13.5.2　TRAFFIC MANAGEMENT MEASURES

These are typically no-entry and banned turn controls which allow buses (and possibly also cyclists) to make a movement prohibited to other vehicles.

More general measures may assist bus movement along a corridor. These measures can be side road closures, banned turns into or out of side roads, one-way streets and yellow-box markings to prevent junctions becoming blocked.

13.5.3　PARKING MANAGEMENT MEASURES

Kerbside parking is a major cause of delays to buses. Parked vehicles make it difficult for buses to approach and leave bus stops and make it necessary for buses to change lane between stops. It is often impractical to ban kerbside parking in urban areas over the whole length of a road at all times of day, because of the lack of rear access to commercial properties requiring deliveries and the lack of off-street parking for residential properties.

Where finance and space permit, one solution is for the highway authority to enable provision of a rear access route to commercial properties and to provide dropped kerbs and allow residents to gain access over the footway to parking within the curtilage of their properties. When this solution is impractical, then it will be necessary to allow waiting and loading at certain times of day and at certain locations. Waiting and loading should be prohibited during bus operational hours at bus stops and on the approach approximately 50 metres prior to the stop line at junctions, which are signal controlled. At other locations waiting and loading may be permitted where essential and during hours when traffic is not congested.

In narrow shopping streets it will often be necessary to prohibit waiting and loading on both sides of the street. One solution is to provide marked bays on the roadway, staggered on alternate sides of the street. Where space permits, consideration should be given to providing bays in footway areas.

13.6 Traffic signal control

13.6.1 SIGNAL TIMINGS FOR PASSIVE BUS PRIORITY

In passive priority the known volume of buses is used to alter fixed-time plan settings or SCOOT parameters. Both BUS TRANSYT and SCOOT can value buses more highly than other traffic and so give a small measure of priority to buses at intersections where other approaches have no bus flow. At many signalised intersections, all approaches are bus routes and so this method is then not effective. In addition, fixed-time co-ordination has to assume that buses spend a constant time at bus stops.

13.6.2 SELECTIVE VEHICLE DETECTION – ACTIVE BUS PRIORITY

In active selective vehicle detection (SVD) the presence of a bus approaching a signalled junction is detected normally by placing a detector some 60–100 metres prior to the stop line. The presence of the bus then either extends the existing phase if it is currently green for the approaching bus or brings forward a green phase. The bus must be detected after any variable delay, such as at a bus stop, and this may require a trade-off with relocation of bus stops away from the optimum position for passengers. SCOOT version 3.1 includes logic to include active bus priority within a SCOOT network.

SVD can provide reductions in delay to buses at isolated signals of around 10 seconds per bus and in SCOOT a reduction of around 5 seconds per bus.

13.6.3 OVERLAP PHASES

A general concern when introducing with-flow bus lanes is that right-turn traffic would queue back beyond the set back to the bus lane. Traffic queuing adjacent to the bus lane would then be blocked and would not be able to flow into the inner kerbside lane and as a result approach capacity would be considerably reduced. One way of overcoming this difficulty is to make use of spare capacity in the direction opposite to peak flow and to provide an overlap phase for right-turn traffic in the peak-flow direction.

The overlap phase for right-turn traffic can either be provided by a late release of the opposing flow (early start of the right turn) or an early cut-off of the opposing flow as in Figure 13.4. Research has shown that accident levels and queue lengths can be lower if late release is employed. One issue that needs consideration with late-release phases is whether to use green-arrow aspects on signal heads, as in Manchester and the North West and how to cancel them.

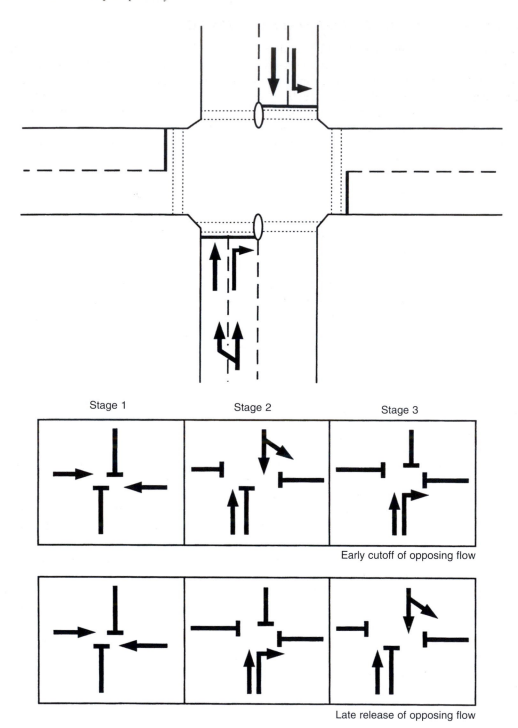

Fig. 13.4 Overlap phase.

13.6.4   QUEUE RELOCATION AND TRAFFIC METERING

In narrow or environmentally sensitive sections of a route it can be impossible or undesirable environmentally to provide bus priority measures such as with-flow bus lanes. One solution is to relocate the queues that form in these narrow sections to wider sections on the same route. Signal control at the end of the wider section can be adjusted to meter traffic into the narrower section so that queues do not occur there. Instead queues are relocated prior to the signals over the wider road section where a bus lane can be installed. This metering of traffic is made easier in SCOOT version 2.4, which can respond to queues detected remotely from the section of road controlled by the immediate signals.

13.6.5   PRE-SIGNALS AND BUS ADVANCE AREAS

With-flow bus lanes are typically set back from a signalled stop lane so that approach capacity is not reduced. One way of providing further advantage to buses, particularly those wishing to make right turns, is to provide a second set of signals, the 'pre-signals' prior to the primary signals. The bus lane is taken to the pre-signals stop line.

While the pre-signals are red to general traffic, buses may be unsignalled or get a green phase and can move to the primary signals stop line using the area between the two sets of signals, the 'bus advance area'. The design of the bus advance area requires a carriageway wide enough for lanes to provide capacity for general traffic, a bus lane and a splitter island between the bus lane and general traffic lane(s) for mounting a signal post and head with aspects for the general traffic.

13.7   Bus stop improvements

13.7.1   BUS STOP CLEARWAYS

One of the major causes of delay in urban areas for buses and general traffic is inconsiderate parking near bus stops. Buses have difficulty in gaining access to bus stops and subsequently rejoining the traffic stream. Passengers have to resort to walking between parked cars to board a bus. With the introduction of modern low floor buses that can carry passengers in wheelchairs, it is particularly important that buses can deploy their wheelchair ramps to bridge the gap between the bus and the kerb Plates 13.3 and 13.4. Bus stop clearways markings should be introduced to help to prevent obstructive waiting and loading at bus stops during the period of operation services. Many new buses are built to the European standard length of 12 metres, to ensure that they can approach the kerb closely enough to deploy the ramp the cage should be designed to a length of 37 metres. Where 18-metre articulated vehicles are used the cage length should be 43 metres (Plate 13.5).

Additional benefits of the bus stopping close to the kerb in a properly designed bus cage are: ease of access for the elderly, the young, heavily laden passengers and parents with children in pushchairs. Overall bus journey time can be improved by quicker boarding and alighting for all passengers.

13.7.2   BUS BOARDERS

Sometimes the introduction of bus stop clearways or waiting and loading restrictions is inappropriate as it uses excessive lengths of carriageway that is under pressure for parking by local residents or servicing for business and commercial premises. In these circumstances and where parking enforcement is not effective buses can be prevented from reaching the kerb. An effective solution is to construct a bus boarder. Bus boarders are local extensions of the footway into the carriageway. The outer edge of full width boarders should be in line with the parked vehicles (i.e. 2.0 metres for car parking, 2.5 metres when adjacent to loading bays). Full width boarders can often be shorter than the length of the buses using the route (i.e. 8.0–9.0 metres for a 12-metre bus) but long enough to allow easy boarding or alighting at all doors. The benefits of boarders are many: buses will stop outside the line of parked vehicles (as before) but will remain for a shorter time as passengers can access the bus quicker, the bus can re-enter the traffic stream easier, space is created on the footway for a shelter and more kerbside space is available for waiting and loading vehicles.

Plate 13.3 Modern bus station Strafford.

Plate 13.4 Modern bus stop.

Plate 13.5 Articulated (bendy) bus.

In some cases it is necessary to use half-width boarders; these have some of the advantages of full width boarders but require more kerbside space and are more vulnerable to illegal waiting and loading.

13.7.3 BUS STOP RELOCATION

The positioning of bus stops is often dictated by practicalities such as avoiding banks and post offices, where space for special deliveries is required, and avoiding creating poor sight lines for side road traffic. Bus stop locations must not unduly block general traffic, and for this reason may need to be provided with half-width (1.5 metres) bus bays which give more space to general traffic while still allowing easy access and egress for buses to and from the main carriageway. If SVD at signals is required, then bus stops need to be ideally at least 70 metres prior to the stop line to maximise the benefits or, alternatively, beyond the exit. Bus stops also need to be positioned to provide easy interchange between routes.

Bus stop relocation must not be undertaken lightly as carefully arranged stop spacing can be disturbed. Poorly considered relocation can lead to excessively long or short distances between stops. Also, some stops will have been placed to provide an interchange between intersecting bus routes or other public transport services. For example, if a stop is moved from a junction approach to the exit side, interchanging passengers might have to cross a busy carriageway to reach the other bus stop.

13.7.4 BUS SHELTERS AND TIMETABLE INFORMATION

Bus shelters are an obvious improvement at bus stops, providing protection from the weather. Timetable information should also be provided at bus stops in common static form listing the schedule of times of buses arriving at the stop and of reaching destinations, or as real-time information.

13.7.5 SPECIAL PROFILE KERBS

Special profile kerbs can be provided to allow bus drivers to align their vehicles precisely at bus stops. These kerbs are designed to minimise tyre damage and be of greater height than standard kerbs. The great advantage is that boarding and alighting passengers have a near-level entry and exit to the vehicle and without the need to step onto the road (Plate 13.6).

Plate 13.6 Bus stop obstructed by loading vehicle.

13.8   Impacts on other users

Bus priority measures should be designed to enhance or protect the needs of all other road users particularly cyclists, pedestrians, people with disabilities and frontagers.

13.8.1   CYCLISTS

Cyclists can benefit from bus priority measures both in terms of ease of travel and of safety; for example, cyclists are allowed to use with-flow bus lanes. This provides a congestion-free route and removes the risks associated with cyclists negotiating parked vehicles, which would not be allowed in bus lanes. Ideally the width of bus lanes are increased at least from 3.0 to 4.3 metres to accommodate cyclists. Where this is not feasible then it will be necessary to accept that bus drivers should be instructed to remain at a safe, non-intimidatory distance behind cyclists during congested periods. When there are no queuing vehicles adjacent to the bus lane, buses can, with care, cross the bus lane line marking to overtake cyclists. It should be remembered that cycle speeds are usually higher than general traffic speeds during congested conditions and are, there-fore, quite acceptable to bus passengers. Cyclists are often permitted to take advantage of other bus priority measures such as contraflow bus lanes, busways, bus gates and pre-signals. Wherever possible bus priority measures should include provisions for cyclists.

13.8.2   PEDESTRIANS

As well as improvements at bus stops, bus priority measures can provide opportunities for improving facilities for pedestrians crossing the road. When a with-flow bus lane is proposed in a road narrower than 11.0 metres it may not be possible to reposition existing pedestrian refuges and so these may need to be replaced typically by pelican or other signalled crossings. Zebra crossings must always be replaced where they cross with-flow bus lanes. The purpose of a bus lane is to allow buses to bypass stationary or slow moving queues of general traffic. There is a real risk to pedestrians stepping into the path of a moving bus from in front of a vehicle in the general traffic lane. Though the Pedestrian Crossing Regulations,[2] backed up by advice in the Highway Code,[3] clearly state that vehicles should not overtake the leading vehicle at a pedestrian crossing, it is often ignored by all drivers, including bus drivers. A signalled crossing provides a positive signal to stop to both buses and general traffic.

The opportunity can be taken at new signalised junctions or altered existing junctions to pro-vide pedestrian crossing stages, normally activated by manual demand.

13.8.3   PEOPLE WITH DISABILITIES

Many bus operators are introducing low floor buses on their services. This will assist people with disabilities in boarding and alighting. The design of bus priority schemes allows for this intro-duction by taking into account the length and swept path of low floor buses.

13.8.4   FRONTAGERS' NEEDS

Providing priority for buses often requires changes to existing waiting and loading restrictions or the imposition of new restrictions. This must be balanced against the legitimate needs of frontagers, reflecting the different requirements of residential, commercial, industrial and retail uses. These needs are normally achieved by providing reasonable time periods when waiting and loading is permitted on the roadway of bus routes. Where this is not possible, then loading spaces can be marked on the roadway, in bays constructed adjacent to the roadway or in spaces on side streets. Alternatively, where footways are sufficiently wide, then footway parking may be appropriate.

13.9   The process of designing and evaluating bus priority measures

Several types of survey are commonly used to assist in the design and evaluation of bus priority measures:

- bus and other vehicle occupancies
- bus journey times
- bus headways
- junction queue lengths and delays
- classified traffic counts
- parking occupancy and duration.

The design process involves consultation with bus operators, local residents, businesses and statutory bodies to identify appropriate and acceptable bus priority measures.

Junction designs and evaluations of time savings are based on the use of modelling programs such as ARCADY, OSCADY, LINSIG and TRANSYT. The assessment approach set out by the Department of Transport[1] is commonly used for the evaluation of scheme benefits.

This evaluation identifies benefits as:

- time cost savings for bus passengers and crew
- bus operating cost savings.

Other benefits often assessed include:

- improved bus service regularity and reliability due to less variation in travel time as the variable delays due to congestion are avoided;
- generated bus patronage due to reduced travel time and improved accessibility for bus passengers;
- time savings enabling the bus operator to keep the same frequency of service using fewer vehicles or to improve the frequency of service with the same number of vehicles;
- decongestion where a shift to bus usage results in less car traffic and so reduces delays for all traffic.

References

1. DoT (1991) *Keeping Buses Moving*, HMSO, London.
2. HM Government (1997) *SI 1997 The Zebra, Pelican and Puffin Pedestrian Crossings Regulations and General Directions 1997*, HMSO, London.
3. DfT *The Highway Code*.
4. DoE (NI) (1997) *Bus Stops A Design Guide for Improved Quality*.
5. LBI (2000) *TfL Bus Stop Layouts for Low Floor Accessibility*, June.
6. DOT (2001) *Traffic Signs Regulations and General Directions*, HMSO, London.
7. TfL (2002) *Bus Stop Layouts for Articulated Buses*, July.

14

Development Process and Sustainable Development

14.1 Planning context

14.1.1 INTRODUCTION

Most developments require planning consent from the local planning authority, that is, district or unitary authority, and the context for obtaining consent is described in legislation such as the Town and Country Planning Act 1990, in government policy guidance and in local authority plans. There are a few types of development which are controlled differently, for example mineral extraction which is determined at a county level, military works and major infrastructure such as strategic roads and airports.

14.1.2 GOVERNMENT GUIDANCE

Government policy guidance is contained in Planning Policy Guidance (PPG) and Regional Policy Guidance (RPG) notes published by the Government. There are currently some 22 PPG notes giving guidance on specific aspects of planning and some 11 RPG notes giving guidance by region.

PPGs are being replaced by Planning and Policy Statements (PPSs) and RPGs by Regional Spatial Strategies (RSS). Table 14.1 lists the existing notes.

Other sources of official planning guidance available from HMSO include:

● Minerals Planning Guidance Notes
● Derelict Land Grant Advice
● Circulars from Government Departments
● Development Control Policy Notes
● Ministerial Statements.

The PPGs apply to England and Wales. The Scottish and Northern Irish planning systems differs slightly from that in England and Wales and so separate policy guidance applies. PPGs are to be used by local authorities in preparing development plans and in deciding on planning applications.

Table 14.1 PPG and RPG notes

PPG/RPG number	Title of note
PPG 1	General Policy and Principles
PPG 2	Green Belts
PPG 3	Housing
PPG 4	Industrial and Commercial Development and Small Firms
PPG 5	Simplified Planning Zones
PPG 6	Town Centres and Retail Development
PPG 7	The Countryside and the Rural Economy
PPG 8	Telecommunications
PPG 9	Nature Conservation
PPG 12	Development Plans and Regional Planning Guidance
PPG 13	Transport
PPG 14	Development on Unstable Land
PPG 15	Planning and the Historic Environment
PPG 16	Archaeology and Planning
PPG 17	Sport and Recreation
PPG 18	Enforcing Planning Control
PPG 19	Outdoor Advertisement Control
PPG 20	Coastal Planning
PPG 21	Tourism
PPG 22	Renewable Energy
PPG 23	Planning and Pollution Control
PPG 24	Planning and Noise
RPG 1	Strategic Guidance for Tyne and Wear
RPG 2	Strategic Guidance for West Yorkshire
RPG 3	Strategic Guidance for London
RPG 4	Strategic Guidance for Greater Manchester
RPG 5	Strategic Guidance for South Yorkshire
RPG 6	Regional Planning Guidance for East Anglia
RPG 7	Regional Planning Guidance for the Northern Region
RPG 8	Regional Planning Guidance for the East Midlands Region
RPG 9	Regional Planning Guidance for the South East
RPG 9a	The Thames Gateway Planning Framework
RPG 10	Regional Planning Guidance for the South West

14.1.3   DEVELOPMENT PLANS

PPG 1 sets the scene on the planning system and confirms this as 'plan-led'. The clear intention is that proposals for development are to be judged against development plans which have been prepared following a statutory process of public consultation. These development plans include:

- *Structure Plans*: essentially set out strategic policies typically over a County Council area.
- *Local Plans*: prepared by District Councils.

- *Waste and Minerals Local Plans*: prepared for a County area or National Park area.
- *Unitary Development Plans*: prepared by Unitary or Metropolitan Authorities (e.g. London Boroughs).

Local Plans should conform to strategic guidance in Structure Plans which should themselves conform to Regional Planning guidance. Unitary Development Plans, which essentially combine Local and Structure Plans, should also conform to Regional Planning Guidance.

Responsibility for transport issues rests with the highway authority, being the County or Unitary Authority, but Local Plans will be prepared by the planning authority, that is, the District Councils, with guidance from the highway authority.

Proposals for development normally need to be in conformity with these plans. Proposals then will often involve, first, gaining acceptance in a Local or a Unitary Development Plan, and second, gaining planning consent.

Transport engineers are typically asked to assist in both these processes and sometimes also in making submissions to the Structure Plan consultation process if the development is of strategic importance.

14.1.4   PPG 6 AND PPG 13

The transport engineer will need to refer to the PPG and RPG notes and other document sources. Two PPGs are in common use by engineers – PPG 6 and PPG 13.

PPG 13 advises local authorities on how to integrate land-use policies and transport programmes in line with the Government's Sustainable Development Strategy. The aim is to:

- reduce growth in the length and number of motorised journeys;
- encourage alternative means of travel which have less environmental impact;
- reduce reliance on the private car.

PPG 6 advises local authorities on how to promote development in town centres, to select sites for development for retail, employment and leisure and to assess retail development proposals. Some of the aims are to:

- locate major generators of travel in existing centres;
- strengthen existing centres;
- maintain and improve choice for people to walk, cycle or catch public transport;
- ensure an appropriate supply of parking for shopping and leisure trips.

14.2   Planning reforms

Planning reforms are changing the existing set of Development Plans.[9]

The Planning and Compulsory Purchase Bill, which spells out the reforms, is expected to come into force in October 2004. The Bill fundamentally sweeps away the current system of county structure and unitary and local development plans in favour of RSS and Local Development Documents (LDD). Importantly, the RSS including Regional Transport Strategy (RTS) will become part of the 'Development Plan' which carries legal weight in consideration of planning applications. This means that local authorities' LDD will have to comply with the regional RSS and its transport strategy.

A statement of core policies and action plans for key areas of change will make up local authorities' LDD with supplementary planning documents, a statement of community involvement and a proposals map. The map will include all transport networks including walking and cycling routes, interchanges and areas subject to demand management proposals.

The main aim is to make planning more efficient while increasing community involvement in the process. The Bill will 'increase the predictability of planning decisions and speed up the handling of major infrastructure projects'.

In addition to changes to plan making, the Bill proposes several improvements to development control. Local authority power to compulsory purchase will be widened, providing it leads to economic, social and environmental benefit. A new scheme of loss payment is proposed, to supplement the value of property and disturbance costs. Given typical opposition to some recent highway schemes, this measure of compensation should, it is hoped, give a boost to promotion and delivery of new transport infrastructure.

Exactly what is proposed and should happen is set out in a series of consultation papers, including PPS 11 and PPS 12 which deal with regional and local planning respectively. The first makes clear the importance of transport to regional spatial planning and a significant innovation of the second insists planning authorities develop a strategic approach to infrastructure.

14.3 The role of the transport engineer

14.3.1 SCOPE OF ADVICE

The transport engineer will assist in the development process by advising on transport issues associated with the development, such as:

- appropriate location and type of development
- access by all modes of transport
- impact on the transport network
- environmental impact
- transport proposals
- parking.

The engineer may be working for the planning or highway authority responsible for preparing development plans and determining planning applications. He/she may be working for a developer in making representations on development plans or assisting in planning applications and in implementing the development. Alternatively, he/she may be working for a third-party in opposing development proposals in draft development plans and planning applications.

14.3.2 APPROPRIATE LOCATION AND TYPE OF DEVELOPMENT

While the choice of location and type of development is primarily based on planning and financial issues, the transport engineer can advise on transport issues involved in sustainable development. For example, the choice of location can result in marked differences in accessibility by alternate modes of transport and in the amount of motorised journeys. The type of development on a given site can similarly affect the amount of travel; for example, a mixed land-use development or development of housing near an existing shopping or employment centre can often be shown to be more sustainable in transport terms.

In assisting in this choice, the transport engineer will be directly assessing the ability of developments to achieve the aims of PPG 13. Accessibility profiles can be of assistance in determining locational policies designed to reduce the need for travel by car.

An example of an accessibility profile, in the form of weighted accessibility indices, is shown in Table 14.2. These indices show average times to reach employment zones in Basingstoke by public transport from a set of future housing developments, both those proposed in the draft local

Table 14.2 Public transport weighted journey times from housing areas to employment areas (min)

Weighting (per 1000 employees)	8.028	6.150	3.150	8.992	8.550	16.000		
Housing development areas	Employment areas						Weighted average	Rank
	Daneshill	Town centre	Viables	Hampshire international park	Houndmills	Commercial centre		
1 Taylors farm	28.0	20.0	37.0	10.0	31.0	25.0	24	1
17 Nth of popley	32.0	32.0	51.0	28.0	22.0	38.0	33	3=
19 Huish lane	35.0	38.0	57.0	63.0	53.0	48.0	49	8
20 E of riverdene	32.0	17.0	34.0	42.0	32.0	17.0	27	2
21 Beggarwood lane	56.0	41.0	37.0	66.0	60.0	51.0	54	11+
22 Kempshott lane	52.0	37.0	37.0	62.0	60.0	47.0	51	10
23 Old Kempshott lane	48.0	34.0	53.0	60.0	43.0	44.0	48	7
24 Park prewett	37.0	30.0	49.0	37.0	25.0	40.0	36	6
30 Wimpey/taywood Homes, bramley	49.0	23.0	40.0	44.0	34.0	28.0	35	5
31 North popley fields	31.0	31.0	50.0	27.0	31.0	37.0	33	3=
32 Hodd's farm	36.0	39.0	58.0	64.0	54.0	49.0	50	9
33 Saunder's/hounsome	56.0	41.0	37.0	66.0	60.0	51.0	54	11=

plan and others suggested by developers. The average times have been weighted by the forecast numbers of employees in each zone.

Incorporating accessibility profiles within best practice should encourage developers to improve the profile of their site by contributing to public transport improvements. In the case of the Taylors Farm development in Basingstoke, the developer would contribute to a new rail station at Chineham on the Reading–Basingstoke line.

Another approach is to make use of a geographic information system (GIS) such as that originally developed for the London Borough of Croydon and shown in Figure 14.1 and now enhanced to form ACCESSION. This will provide information on the accessibility of public transport to residents and also the relative accessibility from residential areas to attractors, for example employment areas and shopping centres. Census information is extracted and road network information taken from the Ordnance Survey's software products defining road network properties.

The location of all bus stops has been added from London Transport BODS (Bus passenger Origin–Destination Survey) to allow the precise calculation of walk distances. Network travel data can be imported from a traffic model such as that provided by the CUBE or SATURN software referenced in Chapter 4.

Besides Accessibility information, the GIS can be used to produce origin–destination trip matrices, route travel times, isochrones and contours.

14.3.3   ACCESS BY ALL MODES OF TRANSPORT

In small developments, the only transport issues may be those of appropriate access to/from the highway network and off-highway parking. Requirements for these are normally set out by the local authority in published documents. Another useful compendium of information is provided by the Institution of Highways and Transportation (IHT).[1]

A new access must be located to be at least a minimum distance from other accesses. This distance depends on the role of the highway providing the access in the highway hierarchy, for example, access road, local distributor, district distributor or primary road. The location and layout of the access must be such that it provides adequate visibility commensurate with safe stopping distances. Figure 14.2 shows the visibility requirements for a two-way straight road and for a driver eye level of 1.05 metres.

The minor road distance is taken as 4.5 metres but can be reduced to a minimum of 2.0 metres for single dwellings or a small *cul-de-sac*. In Figure 14.2, Table A is used for calculating the major road distance when speed measurements on the major road are available; otherwise Table B is used based on the speed limit that applies.

To protect pedestrians, there is an additional requirement for lightly used accesses as shown with a 2.0 metres visibility splay at a reduced eye level of 0.6 metres for children.

In larger developments the form of access will also depend on capacity, environmental and road safety issues. Accesses may well be provided separately for different modes of transport.

14.4   Traffic Impact Assessment/Transport Assessment

14.4.1   THRESHOLDS

Larger developments require an Impact Assessment provided by a transport engineer. There are no statutory rules as to when a development is sufficiently large that it requires this Transport Assessment (TA), but the IHT document *'Guidelines for Traffic Impact Assessment'*[8] suggests thresholds based on either traffic flows or scale of development. The IHT Guidelines concentrate on highway aspects and treat other modes as of secondary importance. The Guidelines provide

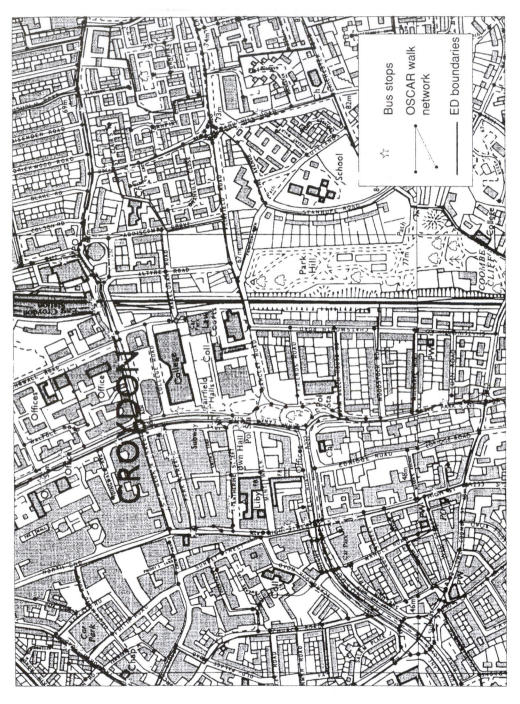

Fig.14.1 Presentation by a GIS.

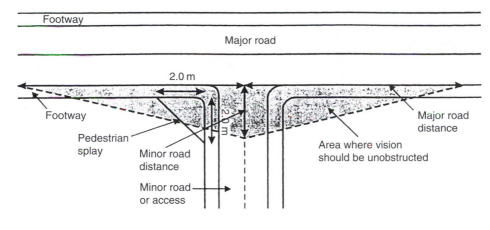

Table A When speed measurements are available

Major road speed (kph)*	120	100	85	70	60	50	40	30
Major road distance (m)	295	215	160	120	90	70	45	33

* 85th percentile speed, i.e. fastest speed excluding the fastest 15% of vehicles.

Table B When speed measurements are not available

Speed limit (kph)	70	60	50	40	30	20
Major road distance (m)	295	215	160	120	90*	45*

* Includes an allowance for motorists travelling at 10 kph above the speed limit.

Fig. 14.2 Visibility requirements.

Table 14.3 TIAs and TAs compared

	TIA	TA
Modes considered	Focus on car but others may be included	All modes considered but emphasis on walk, cycle and public transport
Transport implications covered	Comparison to similar development	Accessibility and mode split analysis
Impacts covered	Road safety and traffic	Wide criteria based on National Approach to Transport Appraisal (NATA)
How are negative impacts addressed	Increased road capacity and add safety features	Wider assessment. Travel Plans. Financial Incentives. Additional road capacity if required

very useful information on traffic impact analysis but do not comprehensively cover aspects that are needed for a TA.

TAs are now required in the current version of PPG 13 and replace Traffic Impact Assessments (TIAs) by including a consideration of access to a development by all modes of transport. The Department for Transport (DfT) is expected shortly to provide guidance on how to prepare a TA. A key consideration in a TA is an analysis of Accessibility and the DfT has prepared guidance on how this should be undertaken using ACCESSION, the software package designed for the purpose.[23]

The key differences between TAs and TIAs are shown in Table 14.3.

The suggested traffic flow threshold for a TIA is when traffic to or from the development exceeds 10% of the two-way traffic flow on the adjoining highway. If this highway is or will become congested in the assessment period (typically 10 years) then the IHT suggest a lower threshold of 5%. An even lower threshold is likely to be required for developments affecting trunk roads.

An alternative approach is to apply thresholds either by size of development for the different Use Classes shown in Table 14.4 or by traffic generation or parking criteria:

- Residential development (C3) in excess of 200 units.
- Business (B1 and B2) gross floor area (GFA) in excess of 5000 square metres.
- Warehousing (B8) GFA in excess of 10 000 square metres.
- Retail (A1) GFA in excess of 1000 square metres.
- 100 trips in/out combined in the peak hour.
- 100 on-site parking spaces.
- Landfill sites or quarries generating heavy goods vehicles (HGVs).

14.4.2   SCOPING STUDY

A key element of a TA is the Scoping Study. The transport engineer should always aim to reach agreement between developer and highway authority on this prior to the assessment work commencing. Without this agreement the TA is likely to prove unsuitable or will require time-consuming modification.

The Scoping Study will include:

- size and nature of development;
- size of Study Area and its network;
- other transport and development proposals that need to be taken into consideration;
- transport data and models available and need for further surveys;
- assumptions on network traffic growth;
- planning standards and policies that apply;
- assumed year of opening of development and its phasing;
- years for assessment.

The size and nature of the development is likely to be well defined by the time a TA is prepared as the definition will be included in the planning application which the TA will support. In a feasibility study for a development, the size and nature of the development may not be defined, but it may still be appropriate to have the scope of the study agreed between developer and highway authority.

The size of the Study Area and its network will depend on the expected traffic generated by the development and current congestion on the network. It may be necessary to include congested junctions remote from the network. The size may also depend on other transport and development proposals known to the highway authority.

Table 14.4 Guide to Use Classes order

Use Classes order 1987	Description
A1	Shops, retail warehouses, hairdressers, undertakers, travel and ticket agencies, post offices, etc.
	Pet shops, cats-meat shops, tripe shops, sandwich bars
	Showrooms, domestic hire shops, funeral directors
A2	Banks, building societies, estate and employment agencies
	Professional and financial services, betting offices
A3	Restaurants, pubs, snack bars, cafés, wine bars, shops for sale of hot food
	Shops selling and/or displaying motor vehicles
	Launderettes, dry cleaners, taxi businesses, amusement centres
B1	(a) Offices, not within A2
	(b) Research and development, studios, laboratories, high tech
	(c) Light industry
B2	General industry
B3–7	Special industrial groups
B8	Wholesale warehouses, repositories
C1	Hotels, boarding and guest houses, hostels
C2	Residential schools and colleges
	Hospitals and convalescent/nursing homes
C3	Dwellings, small businesses at home, communal housing of elderly and handicapped
D1	Places of worship, church halls
	Clinics, health centres, creches, day nurseries, consulting rooms
	Museums, public halls, libraries, art galleries, exhibition halls
	Non-residential education and training centres
D2	Cinemas, music and concert halls
	Dance, sports halls, baths, skating rinks, gymnasiums
	Other indoor and outdoor sports and leisure uses, bingo halls, casinos
	Theatres

It is important to establish the availability of existing traffic data and information on other development proposals, for example, TAs prepared for other developers. Existing transport models can provide useful information and may be required to be used in assessing future transport movements. The need for further surveys for highway traffic movements, public transport patronage and walk/cycle flows should be established. Surveys of journey times and queue lengths may also be required.

Ideally, historic traffic data will provide information on traffic growth. Alternatively, network traffic growth, that is growth that will occur without the development, can be assessed using the Department of Transport's National Trip-End model. Allowance should be made for the lower growth in traffic at peak periods, mainly due to capacity constraint, compared with the growth in daily traffic.

A clarification of the planning standards and policies that will apply both now and in the future is required as these can be in the process of being changed as local authorities prepare new development plans.

Phasing of a major development may be an important consideration as it can determine years for assessment of traffic and also trigger requirements for transport improvements. Assessments are normally undertaken for the year of opening of the development (or years of opening of major phases) and for a future year usually 10 or 15 years after opening.

14.5   Elements of a TA

The TA should cover the following:

- existing conditions
- proposed development
- trip generation, distribution, modal split and assignment
- transport impact
- transport proposals
- environmental impact
- accessibility analysis.

14.5.1   EXISTING CONDITIONS

Site visits will identify the nature and condition of the current transport network. The information to be collected will include road geometric data and junction layout, bus and rail services, and pedestrian and cyclist facilities. The visits will also assist in assessing current transport problems, for example highway congestion and lack of adequate public transport service.

Existing traffic data will be identified and analysed and the results of the further surveys initiated as part of the TA will be described.

Accident data may be analysed and any local accident problems identified.

Other developments with planning consent but not yet implemented and planned transport improvements should be listed.

14.5.2   PROPOSED DEVELOPMENT

The proposed development will be described including:

- current or former use of site and any relocation proposals;
- planning policies applying to this site;
- any previous planning applications;
- site area and boundary;
- scale and size of land-use proposed;
- hours of operation;
- development phasing;
- site plan for development.

14.5.3   TRIP GENERATION, DISTRIBUTION, MODAL SPLIT AND ASSIGNMENT

Techniques for estimating travel demand have previously been described in Chapter 4.

There is now a considerable amount of information on the amount of highway traffic generated by new developments. The TRICS database[2] is one well-established national database which holds survey information from hundreds of sites covering many different land uses. These land uses include retail superstores, retail parks, warehouses, offices, business parks, industry, leisure and housing, often with sub-categories for each usage. The TRAVL database provides more detailed information for London sites including the geographic distribution of journeys.

Table 14.5 Typical trip generation rates

Land use	Peak arrivals (a.m.)	Peak departures (a.m.)	Peak arrivals (p.m.)	Peak departures (p.m.)	Total daily arrivals	Total daily departures
Offices, trips/100 m^2	1.5	0.1	0.1	1.1	4.8	4.8
Business parks, trips/100 m^2	1.2	0.1	0.2	0.9	4.0	4.0
Warehousing, trips/100 m^2	0.3	0.1	0.1	0.3	2.1	2.1
Residential, trips/household	0.2	0.5	0.5	0.2	3.9	3.9
Industrial, trips/100 m^2	0.7	0.2	0.2	0.6	4.2	4.2
Hotels, trips/bedroom	0.2	0.2	0.2	0.2	3.2	3.2
Retail parks, trips/100 m^2	0.5	0.2	0.8	1.0	12.2	12.2
Supermarkets, trips/100 m^2	2.4	0.7	6.2	6.4	68.0	68.0

TRAVL includes information on other modes of travel (i.e. public transport, walk and cycle) generated by developments and TRICS has now commenced collecting this. It is often necessary to resort to more general sources of travel information, such as the National Travel Survey,[3] for detailed information on journeys from housing areas for each journey purpose.

When using information from the existing databases or other sources then the approach to estimating trip generation is one of comparison. The development proposal is compared with similar developments for which traffic data are available on the basis of floor area, employment or other appropriate variable.

If suitable comparative information is not available, then trip generation can be assessed either from first principles by making broad assumptions on travel behaviour or by surveying travel movements at a similar site. This latter method is particularly important in countries outside the UK where comparative data may be unavailable.

Typical trip generation rates are shown in Table 14.5 for peak hour and total daily vehicular traffic.

The estimation of distribution and modal split can be achieved using a full transport model as described in Chapter 4 or by a simpler process either based on existing traffic movement data or using isochrones or gravity distribution assumptions.

Existing traffic movement data can be obtained from a survey of movements at an existing similar nearby land use. Alternatively, Census journey to work tabulations may provide a guide.

With a gravity model (see Section 4.4), population and employment data and information on journey time or distance is input to the model formula to estimate the distribution of trips to/from a generating development over the surrounding area. The form of the deterrence function is sometimes simplified to either (travel time)^2 or (travel distance)^2 if information on travel cost is not available, so that:

$$T_{ij} = P_j G_i / C_{ij}^2$$

where T_{ij} is the trips between origin zone i and destination zone j,

C_{ij} is the travel time between zones i and j.

P_j is the population of destination zone j,

G_i is the trips generated by the proposed development in origin zone i,

α has a balancing value to ensure that

$$\sum_i T_{ij} = G_i$$

An alternative, even simpler approach is to establish isochrones of equal travel time from the development. By assuming a trip length distribution (i.e. number of trips within each time band) from a similar site, the number of trips can be distributed into each time band using the isochrones.

When considering retail developments it is often desirable to make use of a shopping model which includes other retail stores and assesses demand for all competing stores. The output from such a model will include the forecast distribution of journeys to the development.

The forecast of trips generated by a retail development will also include an analysis of 'pass-by' and 'diverted' trips being those journeys that already exist on the road network and will be broken to include a visit to the development.

If a sophisticated model is not available then modal split will have to be estimated from existing data surveys, for example observations from similar sites or from Census data.

Trip assignment models were described in Chapter 4. In TA work, manual assignment is often more appropriate than using a computer model as the number of zones is sufficiently small for a manual approach to be practical.

14.5.4 TRANSPORT IMPACT

The transport impact conventionally concentrates on highway link and junction analysis but should also include road safety and bus and rail demand.

Link flows after development should be compared with values in Department of Transport, DoT TA46/97 'Traffic Flow Ranges for the Assessment of New Rural Roads'. This will provide an assessment of the operational performance of road links. Junctions are conventionally assessed using traffic simulation programs, such as ARCADY, PICADY, OSCADY, LINSIG and TRANSYT. Merging, diverging and weaving performance on and between slip roads may also need checking.

The highway analysis will normally be summarised as a set of reserve capacities and queue lengths at critical locations for the assessment years appropriate to the development.

The likely impact of the development traffic on road safety will be assessed, based on the examination of historic accident data.

The impact of demand on existing bus and rail services may also need to be assessed to determine any shortcomings in existing provision or any overloading that is likely to occur with the development.

The requirement for parking and any impact on neighbouring streets will be assessed.

14.5.5 TRANSPORT PROPOSALS

Transport proposals will be put forward to mitigate any negative transport and environmental impacts likely to be caused by the development.

These proposals will include improvements to:

- accesses
- highway junctions and links
- pedestrian and cyclist routes, and provision

- public transport services, priority and facilities
- facilities for the disabled
- parking
- road safety
- internal road layout.

The proposals will themselves be assessed to measure their performance in mitigating impacts or in providing improvements affecting existing residents and businesses.

14.5.6   ENVIRONMENTAL IMPACT

This is covered in more detail in the next section but will include:

- noise
- vibration
- community effects (e.g. severance)
- air quality.

14.6   Accessibility planning using ACCESSION

ACCESSION is the new software provided by MVA to the DfT for use nationally in Accessibility Planning Activities (Figure 14.3).

It enables both the public and private sector to provide solutions which help tackle social exclusion, in development planning and in improving access by public transport.

Acquisition, preparation and application of appropriate transport, land use and socio-economic data are essential if accessibility analysis and audit is to be reliable. The information must also be consistent, sufficiently accurate and up to date.

To undertake an accessibility audit one needs to incorporate public transport datasets of bus stops and routeing data (in formats such as ATCO CIF), Census information and key origins and destinations in your area.

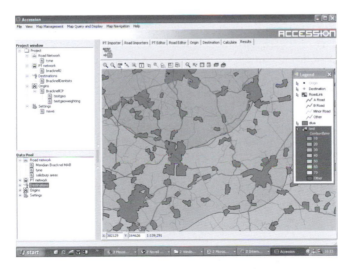

Fig. 14.3 ACCESSION User Interface.

Accessibility Audits provide information on travel times by different modes (e.g. car, public transport, cycle and walk) and use GIS-based mapping to relate this to data on population, employment and a range of social criteria.

Local Authorities need Accessibility Audits to prepare action plans to tackle social exclusion, as an essential input to Development Frameworks and Local Transport Plans and as a tool for development control.

Health and Education service providers can make use of Accessibility Audits to plan the location of services and provide transport where access to services from areas of need is poor.

Public Transport Operators will find Accessibility Audits a useful tool in providing services in response to objectives and plans set by local authorities.

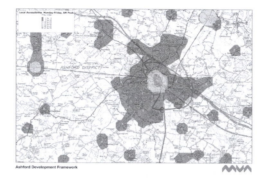

Fig. 14.4 Ashford Local Accessibility.

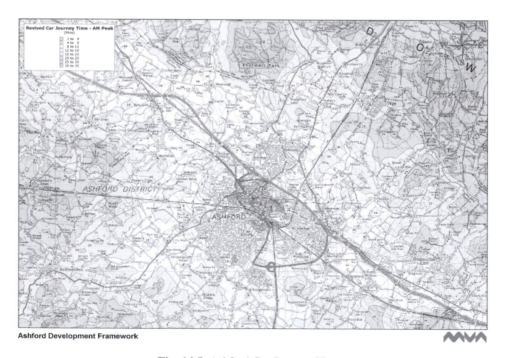

Fig. 14.5 Ashford Car Journey Times.

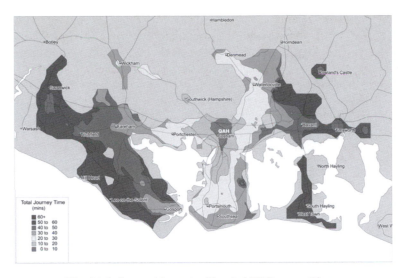

Fig. 14.6 Queen Alexandra Hospital PT Journey Times.

For the Development Sector, Accessibility Audits will provide information for developers, landowners and their consultants to identify locations suitable for specific types of land-use and appropriate development mixes at individual sites. The Audits also indicate where transport improvements are required to render a site viable for a particular development.

The Audits will be an essential element in the preparation of TAs, Environmental Impact Assessments and Travel Plans.

Illustrated examples include:

- LB Harrow for planning magistrate services (Figure 14.8),
- Ashford Borough Council for town centre planning (Figures 14.4 and 14.5),
- Portsmouth NHS Trust for site and service development planning (Figure 14.6),
- Bovis and Heron Homes for preparing sustainable transport proposals for a housing development (Figure 14.7),
- BT for managing travel and parking and improving access by all travel modes to their Adastral Park site (Figure 14.9).

PPG 3 states that Transport Assessments '*should illustrate accessibility to the site by all modes and the likely modal split of journeys to and from the site*'. The Guidance then describes the requirement for analysing accessibility in more detail, as follows:

> *A key planning objective is to ensure that jobs, shopping, leisure facilities and services are accessible by public transport, walking and cycling. This is important for all, but especially for those who do not have regular use of a car, and to promote social inclusion. In preparing their development plans, local authorities should give particular emphasis to accessibility in identifying the preferred areas and sites where such land uses should be located, to ensure they will offer realistic, safe and easy access by a range of transport modes, and not exclusively by car.*

Accessibility analysis is also included in the DfT draft document 'Transport Assessments: A Good Practice Guide'.

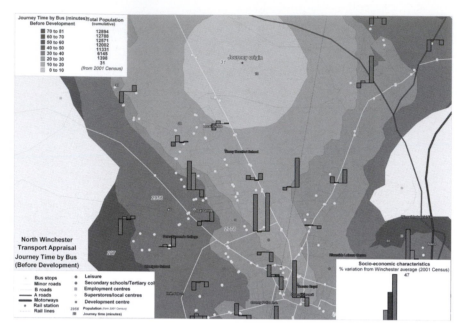

Fig. 14.7 North Winchester Journey Times by Bus.

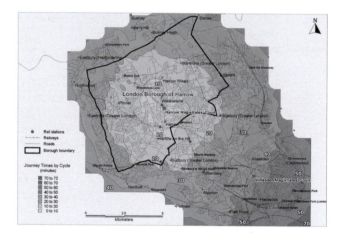

Fig. 14.8 LB Harrow Journey Times to Magistrate's Court.

There are two ways of measuring the accessibility of a location:

- **Local accessibility**, identifying how accessible a location is to public transport services by measuring walk and wait times, often by using the PTAL index (1–6) pioneered by the London Borough of Hammersmith and Fulham.
- **Network accessibility**, measuring journey times or costs across the transport network to the location from all other nearby areas separately for each mode (e.g. car, public transport, cycle and walk).

Network accessibility is more useful for development planning.

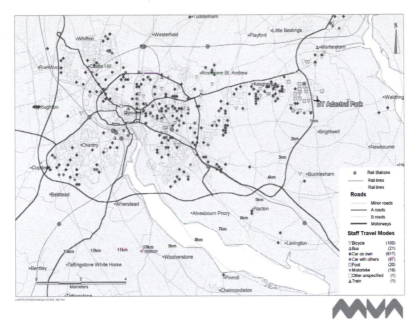

Fig. 14.9 Adastral Park Staff Travel Modes.

Journey times or costs are calculated and then displayed as isochrones (time contours) or cost contours.

The advantage of using GIS software, such as ACCESSION, is that more detailed accessibility information can then be extracted from the mapping using geocoded land-use statistics. For example, employment opportunities within a set travel time from the location, perhaps 30 minutes, can be assessed for each travel mode. This can be used to assess the relative merits of competing locations for housing development.

In addition, the effect of public transport improvements in improving access from a housing development location to employment opportunities or local services (e.g. schools, shops and hospitals) can be assessed.

Accessibility for types of development that are destinations (e.g. office, retail, health and education) can be evaluated by determining the population within a catchment area corresponding to a defined travel time.

A further useful indicator of the relative accessibility and, hence, the sustainability of alternative locations for new development is accessibility ratio. This is the ratio of journey times by car and public transport and is normally calculated by summing journey times to other areas weighted by the 'value' of those areas. An accessibility ratio of a proposed housing location would have journey times weighted by the number of employees in these areas.

14.7   Environmental assessment

14.7.1   FORMAL REQUIREMENTS

The need for formal Environmental Assessments (EA) was set out by the Department of the Environment (DoE).[4] This followed implementation of the European Community Directive No. 85/337 in the 1988 UK Town and Country Planning Regulations.

EA is required if the particular development proposal is likely to have significant effects on the environment. Some guidance is given on thresholds for different types of development needing EA. Those most likely to affect traffic engineers are:

(a) *Manufacturing industry*: sites > 20 hectare.
(b) *Industrial estates*: sites > 20 hectare or with more than 1000 dwellings within 200 metres of the site boundary.
(c) *Urban developments*: sites > 5 hectare or with more than 700 dwellings within 200 metres of the site boundary or providing more than 10 000 square metres gross floorspace of shops, offices or other commercial uses.
(d) *Urban roads*: where more than 1500 dwellings lie within 100 metres of the centre line of the road.
(e) *Other projects*: sites > 100 hectare.

The EA involves collecting information on the likely environmental effects of the development on human beings and on flora, fauna, soil, water, air, climate, landscape, material assets and cultural heritage.

The assessment includes indirect effects such as those caused by traffic.

The planning authority will advise whether an EA is required. If it is required then an Environmental Statement containing the information collected during the EA is to be provided with the planning application.

14.7.2   ENVIRONMENTAL EFFECTS CAUSED BY TRAFFIC

Most developments will not require a formal EA. Even so, there may well be a need for the assessment of environmental effects caused by traffic. The Institute of Environmental Assessment has published its own Guidelines[5] on how this is to be assessed and prepared the following thresholds to identify road links and areas needing assessment:

(a) Road links where traffic flows or numbers of HGVs increase by more than 30% as a result of development traffic.
(b) Sensitive areas (e.g. hospitals), conservation areas or areas of high pedestrian flow, where traffic flows or numbers of HGVs increase by more than 10% as a result of development traffic.

The 30% guideline is approximately equivalent to an increase in noise level at residential frontages of 1 decible (A) which is about the smallest change in level discernible to people.

The Guidelines recommend that assessments be undertaken in the year of opening of the development and over time periods when the environmental impact is greatest; these periods may not coincide with periods of maximum traffic flow.

The items to be measured include:

- noise assessed near buildings;
- vibration in buildings;
- visual obstruction and intrusion;
- community effects such as severance caused by difficulty in crossing by road or by the physical barrier created by the road;
- delays to drivers and pedestrians;
- pedestrian amenity, fear and intimidation;
- accidents and safety;
- hazardous loads;
- air pollution;

- dust and dirt (e.g. quarrying sites);
- ecological effects;
- heritage and conservation areas.

Procedures for measurement are included in the DoT's *Manual on Environmental Assessment* included as *Volume 11* in the *Design Manual for Roads and Bridges*.[6] Further technical background information is available in other documents.[7]

Mitigating effects to minimise environmental impact will include capacity improvements, pedestrian crossings, restriction on movements of HGVs (both hours and routeings) noise barriers and traffic calming.

14.8 Sustainable development

14.8.1 INTRODUCTION

A widely agreed definition of Sustainable Development is 'development that meets the needs of the present generation without compromising the ability of future generations to meet their own needs'.[10]

The Government's Panel on Sustainable Development[11] considers that the Dft's projected growth of road traffic over the next decades is unsustainable. The Panel supports the views of the Royal Commission on Environmental Pollution[12] on the need for clear objectives and quantified targets with an increased emphasis on the role of public transport.

These are hardly new views. Forty years ago, Buchanan[13] had noted that cars are inefficient in use of space, both on the road and at the destination, and that it would be physically impossible to accommodate everyone in the city centre if they were all to travel by car.

It was also clear, by observing the impact on small and medium towns and cities in the USA, what happens to the vitality of town centres if major generators of travel demand are located on the edges of towns. Yet this is precisely what has happened in the UK in the 1980's and 1990's.

PPG 13[14] might then be considered as a classic case of attempting to lock the gate after the horse has bolted.

14.8.2 GOALS AND POLICIES

PPG 13 provides guidance on how local authorities can adopt policies to meet the Government's commitment to a sustainable development strategy. These policies are intended to contribute to the goals of:

- improving urban quality and vitality,
- achieving a healthy rural economy and viable rural communities.

Figure 14.10 shows some of the linkages between the strategy and the adopted policies in summarised form. The implication is that it is essential that land usage and transport needs be planned together to contribute to the sustainable development strategy.

14.8.3 OPPORTUNITIES AND CONSTRAINTS

Planners and engineers will seek the appropriate opportunities to fulfil these policies but will be subject to a number of practical constraints:

- Opportunities:
  - Change land-use and transport patterns.
  - Manage traffic demand.

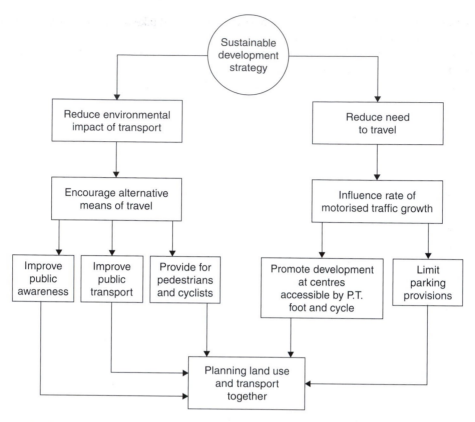

Fig. 14.10 Linkages between sustainable development strategy and adopted policies.

  – Develop transport policies and proposals through the planning process.
  – Contribute to public awareness of the unsustainable consequences of present actions.
● Constraints:
  – Public resistance to change of mode of transport.
  – Demand for choice of mode of transport.
  – Competition between local authorities for development.
  – Need for urban regeneration.

14.8.4   CHANGE LAND-USE AND TRANSPORT PATTERNS

SACTRA[15] observed that road improvements induce additional traffic mainly due to:

  ● the total volume of activities;
  ● the location of activities;
  ● the timing of activities;
  ● the mode of transport used;
  ● the co-ordination of activities by different individuals;
  ● the route chosen.

  By altering the location of activities and by reducing the utility of travel by car, it would be the aim to achieve a *reduced traffic* effect, that is the opposite of the effects observed by SACTRA.

One of the difficulties is that changing land-use patterns will take a very long time because of the relatively slow turnover, 30 years or more, in re-use of land.

Another difficulty is that there are so many out-of-town retail centres that more people now use them rather than town centres for their weekly convenience shopping needs. It can be argued that additional out-of-town retail centres for convenience shopping may actually reduce the need for motorised travel.

14.8.5   MANAGE DEMAND

Some measures, well known to transport planners and engineers, for managing demand for car traffic are:

- improve heavy rail and introduce light rapid transit (LRT);
- improve bus frequency, reliability and quality of service;
- implement bus and LRT vehicle priority;
- reduce public transport fares;
- increase fuel charges for motorists;
- traffic calming and environmental traffic management;
- Park and Ride;
- pedestrianisation schemes/area bans on certain vehicles;
- improved facilities for pedestrians and cyclists;
- improved information on public transport services.

These measures are all part of the toolkit for managing demand, particularly in urban areas. To be effective these measures need to be coupled with either parking restraint or congestion charging.

Parking restraint implies:

- control of both number and price of public on-street and off-street parking;
- limiting the number of spaces allowed in new developments;
- possibly controlling existing private non-residential parking.

Congestion charging would be achieved in towns and cities by one of the forms of urban road pricing or area licensing and between towns and cities by motorway tolling or road pricing.

14.8.6   DEVELOP TRANSPORT POLICIES AND PROPOSALS THROUGH
THE PLANNING PROCESS

PPG 13 makes it clear that the transport policies and proposals should aim to:

- support the locational policies and objectives of the plan (Local, Unitary or Structure);
- improve the environment;
- reduce accidents.

Local authorities drawing up their Transport Plans are expected to demonstrate that their proposals harmonise with the guidance in PPG 13.

PPG 13 also implies that there should be no protection of road corridors for schemes on which work will not commence within the plan period. This may effectively rule out the planning of new roads in urban areas to cater for the long-term traffic growth projected by the DfT.

PPG 13 highlights the desirability of avoiding loading primary roads with short-distance local traffic and of avoiding accesses from new developments directly onto primary roads. Land-use policies should then support transport aims, including safeguarding the role of primary roads.

14.8.7 CONTRIBUTE TO PUBLIC AWARENESS

Transport planners and engineers can contribute to public awareness of the issues involved through initiatives such as Hampshire County Council's HEADSTART. The public in many towns is aware, through personal experience, of the effects of congestion, which have been assessed by the CBI at £15 billion per year in the UK. The environmental effects of transport may not be so obvious; Professor Pearse[16] has suggested that the health cost of particulate emissions from vehicles is around £14 billion per year in the UK.

Short trips in cars are particularly emission inefficient, partly because of the limitations of catalytic converters.[17] If new development is concentrated in existing centres, as recommended by PPG 13, then short trips should increase as a percentage of all trips. It is these short trips, of length <5 miles, which can most readily transfer from car to walk, cycle or bus. Public awareness of the issues could enable this transfer.

The Countryside Commission[18] has also pledged to increase public awareness of the impact of traffic growth upon the Countryside as well as publishing practical guidance on rural traffic calming and demand management.

14.9 Public resistance to sustainable development policies

14.9.1 CAR TRAVEL AND CHANGE OF MODE

The privacy, practicality, quality and convenience of car travel provides a very obvious resistance to change of mode from car to public transport, walk or cycle.

Bus patronage has declined in the UK overall by about 22% between 1985 and 1994. A study of five towns[19] showed that if this lost patronage could be recovered, then buses would carry 16–44% of urban travellers using only 3–11% of the road space. Several improvements are necessary to reverse the decline in bus usage:

- Bus priority to improve reliability and journey times.
- Modern vehicles to provide a high-quality travelling environment and full access for those with disabilities.
- Better information about services.
- Better terminal and bus stop facilities.
- Improved frequency of service.

Some reasons for resistance to change to walk and cycle are:

- lack of safe, direct, well-lit routes and paths for cyclists;
- lack of secure, weather-protected cycle parking;
- lack of safe routes for pedestrians, particularly children.

The latter point is particularly important and has been a major reason for the dramatic reduction over the last 20 years in the percentage of children walking to school. So often highway authorities only improve pedestrian facilities in response to accidents and in line with DfT guidelines rather than as a response to encouraging walking.

Countries such as Germany, Netherlands and Denmark have achieved a greater public acceptance of public transport, walk and cycle modes than the UK. The Royal Commission of Environmental Pollution[12] aims to move the UK in the same direction and recommended the setting of targets to increase public transport use from 12% in 1994 to 30% by 2020 and to increase cycle use of all urban journeys from 2.5% to 10% by 2005; these targets now appear to be unrealistically high.

14.9.2   DEMAND FOR CHOICE

Living in a market economy we expect to have choice and freedom of movement. This is true for seeking employment opportunities, shopping or leisure activities. Pricing or other demand management controls on transport need to be seen by the public as providing benefits that outweigh the disadvantages.

14.9.3   COMPETITION BETWEEN LOCAL AUTHORITIES

Local authorities compete to maintain or improve their economies. This may lead to the offering of incentives to employers to locate in an authority's area by granting employers the numbers of car parking spaces they desire.

It can also lead to authorities setting parking charges in their town centres at a low level or providing further parking spaces to attract shoppers and visitors away from town centres in other authorities. There will inevitably be a conflict between PPG 13 and further provision of parking spaces for shoppers, as advocated by the House of Commons Environment Committee and the Secretary of State for the Environment.

Parking policy can clearly, for these reasons, have repercussions well beyond the boundary of an individual local authority.

14.9.4   NEED FOR URBAN REGENERATION

Many authorities will in practice give the need for urban regeneration and the resultant creation of employment opportunities a higher priority than reducing congestion or reducing pollution. Prospective employers currently often demand sufficient parking space so that all employees can drive to work.

PPG 13 must not be considered to negate the opportunity for urban regeneration.

14.10   Responses by transport planners and engineers

14.10.1   GOVERNMENT POLICY

Transport planners and engineers have witnessed a major shift in Government policy towards surface transport since 1995. PPG 13, while setting out the new stand on transport planning issues, is surely not the Government's last word on the subject.

More fundamentally, the Government Panel on Sustainable Development recognises 'the Government's wish for a broad national debate on the subject followed by a thorough and measured response'. The Government report on road pricing, published in Summer 2004, is moving this debate forward.

14.10.2   SUSTAINABLE DEVELOPMENT MEASURES

Some of the opportunities available for enabling the sustainable development process are:

- local employment and shopping opportunities;
- walkway and cycleway networks;
- roads designed to deter short-distance journeys by car;
- public transport improvements;
- creation of a car-free culture;
- telematics applications.

If employment and shopping opportunities are available close to residential areas then there is an increased likelihood of journeys being contained within the immediate area. This will encourage short-distance journeys. If a network of walkways and cycleways is provided then these short-distance journeys are more likely to be made on foot or by cycle.

In a new housing development much can be done to aid journeys by foot or cycle if the road layout discourages short-distance journeys by car within the development area and to the local centre.

Public transport improvements and locations of developments at nodes well geared by public transport will encourage a higher usage of public transport services. If linked with demand management measures then modal share by public transport can realistically be increased.

A car-free culture can be created by a combination of these measures with a travel awareness campaign.

Telematics applications are discussed in more detail in Chapter 15 but can minimise the need for travel and provide information for more efficient journeys.

14.10.3   PERFORMANCE INDICATORS

The performance indicators that may be used by transport engineers for measuring the sustainability of development proposals include:

- containment of journeys within a development area;
- daily car journeys per household;
- daily car kilometres per household;
- modal share;
- Road Congestion Reference Flows;
- impact on adjoining settlements;
- viability of public transport services;
- accessibility or social inclusion.

14.11   Travel plans

14.11.1   SOFT DEMAND MEASURES: GENERAL

Travel Plans are often considered to be 'soft' measures influencing travel behaviour. In practice they are part of a group of 'soft' measures which include:

- workplace travel plans;
- school travel plans;
- personal travel planning (travel plans for individuals);
- car clubs (clubs providing cars shared between members);
- public transport information (information on public transport schedules and interchanges);
- home/tele working (working for part or all of the week at home);
- videoconferencing.

Workplace and School Travel Plans are prepared by organisations, such as employers, hospitals, universities and schools, to enable them to meet their travel objectives. These objectives will be specific to each organisation but will include the national objectives of reducing the amount of travel in single-occupancy cars, and improving access by all travel modes.

Most hospitals and universities faced with rising parking demand and inadequate availability of land or finances to provide more car parking spaces, have been the quickest to implement Travel

Plans. Schools often implement Travel Plans as 'Safe Routes to School'. Employers have been less enthusiastic to prepare Travel Plans and often only do so when they are required by local authorities as a condition of planning consent. There are notable exceptions, such as Boots in Nottingham, Pfizer in East Kent and Stockley Park in West London, who have implemented Travel Plans for sound commercial reasons.

14.11.2   THE WHITE PAPER

The *Government White Paper*,[20] which leant heavily on advice from Transport 2000,[21] suggested the following benefits of a Travel Plan:

- Strengthens environmental performance and improves environmental image.
- Offers substantial savings by reducing the need for workplace parking and releasing land and buildings for more productive uses.
- Makes work sites less congested and more accessible for deliveries and visitors and improves relations with neighbours.
- Helps staff arrive on time and with less stress by improving travel arrangements.
- Attractive benefits and savings for employees enhance the recruitment package.
- Promotes equal opportunities by providing travel perks throughout the organisation.
- Helps staff to be healthier, fitter and more productive by encouraging exercise.

The emphasis is on the Journey to Work and the contribution that Travel Plans can have in providing alternatives to driving to work alone. The Government envisages implementation of Travel Plans by companies being voluntary but expects local authorities to encourage companies by:

- their own example in developing Travel Plans;
- setting targets in Local Transport Plans;
- entering partnerships arrangements with business and the wider community.

14.11.3   THE TRAVEL PLAN CONTENTS: SOME EXAMPLE

There is now a plethora of advice on how to prepare and implement a Travel Plan. Typically the Plan will contain a number of carrots and sticks to reduce car use for the journey to work. Carrots can include:

- organised arrangements for car sharing;
- improved access by bus, walk and cycle modes;
- better site facilities for cyclists;
- Park and Ride services;
- better information for staff on public transport services;
- financial incentives for cycle and bus travel and for car sharing.

Sticks can include:

- restrictions on permits for staff parking on site;
- reducing car subsidies for employees;
- charging staff for permission to park on site.

Southampton University Hospitals NHS Trust was the trail blazer for implementing many of these carrots and sticks. At its General Hospital site, the Trust wanted to expand its facilities and in doing so estimated that it needed a second multi-storey car park. The City Council objected on the grounds of increased congestion on local streets and environmental disbenefits for local residents.

The Trust responded by implementing a series of measures, achieved a reduction of 600 in the demand for parking spaces and so were able to expand hospital facilities without needing a new car park. The measures included:

- payment of £50 to staff who relinquished their parking permit;
- improved cycle racks and shower facilities;
- free taxi service between hospital sites when on hospital business;
- reduced cost for bus season tickets;
- employees living within one mile of the site or on a bus route serving the site only get parking permits in special circumstances;
- free Park and Ride service from an underused supermarket car park at Lordshill, near a motorway junction;
- information maintained and provided on clusters of staff living near to one another who wish to car share;
- subsidised taxi ride home for car-sharing staff who unexpectedly have to work late and miss their ride.

Examples of measures from other Travel Plans include:

- funding of Travelwise officer to raise awareness among staff of transport issues and encourage a switch away from driving by car to work;
- prominent displays of current public transport timetables;
- personalised public transport timetable for journeys between home and company site;
- new cycle lanes;
- loans to buy a bicycle or obtain season tickets;
- new pedestrian crossings and well-lit pedestrian routes;
- lunch-time shuttle bus to local retail centre;
- company-financed minibus service to local rail station;
- high rate offered to staff for cycle mileage on business.

14.11.4   PREPARING AND IMPLEMENTING A TRAVEL PLAN

Every organisation is different. Yet there is enough commonality for there to be a relatively standard methodology for developing a Plan. Some or all of the following steps will be required:

1. Identify the Organisation's Travel Objectives.
2. Site and Access Surveys and Staff Travel Survey.
3. Staff Group Discussions.
4. Assessment of Accessibility.
5. Forecast of Future Demand for Travel.
6. Identify Options for Measures and Policies.
7. Assess the Benefits and Costs.
8. Consult on Draft Travel Plan.
9. Commit to Implementing the Plan.
10. Implement, Manage and Monitor the Plan.

The Staff Travel Survey questionnaire is normally sent to all members of staff and includes a travel log. The questionnaire is analysed to identify travel behaviour such as mode of travel by home address, need for use of car for business or linked trips (e.g. taking children to school), and scope for changing mode.

Site and access surveys will cover current use of parking spaces compared with capacity, parking permit arrangements for staff and visitors, safe access routes for pedestrians and cyclists, and convenient access to public transport and availability of timetable information on services.

Structured group discussions or meetings of a Staff Transport Working Group can identify staff motivation for choice of travel mode, their perception of local barriers to travel other than by car and their attitude to measures to encourage change in travel behaviour. They will be a prime means of identifying options for measures and policies.

An assessment of accessibility from staff residences to company site by different travel modes can be undertaken using manual or automated network analysis. The assessment can lead to identification of the need for new or improved bus services, cycle and pedestrian routes.

If the use of the site is likely to change, then future demand for journeys to work to/from the site will need to be forecasted.

The costs and effectiveness of each measure and policy in achieving desired benefits (such as reducing car travel) is assessed based on the travel survey results, the comparison of accessibility (e.g. journey times) by different modes from home to site and staff attitudes.

The organisations will need to have top management commitment to the Plan and adequate financial and staff resources to ensure its implementation. Often, the post of a part- or full-time Travel Plan Co-ordinator is needed to provide ongoing management and monitoring of the Plan.

14.11.5 IMPACT OF TRAVEL PLANS

School and Workplace Travel Plans often include other soft measures such as improved public transport information, homeworking and videoconferencing as well as improvements to bus, walk and cycle access and parking controls and charges. Several researchers have prepared estimates based on assessments of the impacts of existing travel plans and these estimates are reported by Goodwin et al.[22] The main findings of their work were that:

(i)   Workplace Travel Plans reduced car use by between 5% and 28% overall with good travel plans achieving average reductions of 18%.
(ii) School Travel Plans reduced car use by up to 50% with good plans achieving reductions in the range 20–30%.

Goodwin's conclusions were that soft measures should have an important role in transport strategy and were very good value for money in reducing car use.

References

1. Institution of Highways and Transportation (1997) *Transport in the Urban Environment*, IHT, London.
2. JMP (undated) *TRICS – A Trip Generation Database for Development Control*, JMP, London.
3. Department of Transport (undated) *National Travel Survey*, HMSO, London.
4. Department of the Environment (1989) *Environmental Assessment – A Guide to the Procedures*, HMSO, London.
5. Institute of Environmental Assessment (1993) *Guidelines for the Environmental Assessment of Road Traffic*, IEA, London.
6. Department of Transport (1993) *Design Manual for Roads and Bridges*, Volume 11, Environmental Assessment, HMSO, London.
7. Morris, P and Therivel, R (1995) *Methods of Environmental Impact Assessment*, University College, London.

8. Institution of Highways and Transportation (1992) *Guidelines for Traffic Impact Assessment*, IHT, London.
9. Martin Shaw (2004) *Opportunity Knocks for Transport Planners*, Transportation Professional, March, IHT, London.
10. Brundtland, World Commission on Environment and Development (1987) *Our Common Future*, Oxford University Press.
11. British Government Panel on Sustainable Development (1995) *First Report*, DoE.
12. Royal Commission on Environmental Pollution (1994) *Eighteenth Report: Transport and the Environment*, HMSO.
13. Buchanan, C (1963) *Traffic in Towns: A Study of the Long-term Problems of Traffic in Urban Areas*, HMSO, London.
14. Departments of the Environment and Transport (2001) *Planning Policy Guidance – Transport (PPG 13)*, HMSO.
15. The Standing Advisory Committee on Trunk Road Assessment (1994) *Trunk Roads and the Generation of Traffic*, HMSO.
16. Baker (1995) *Sustainable Development Balancing Act Waits for Government Lead, Surveyor*, 23 February.
17. Royles (1995) *Literature Review of Short Trips, Project Report 104*, TRL, Crowthorne.
18. *Sustainability and the English Countryside*, Countryside Commission, 1993.
19. *The Role of the Bus in the Urban Economy*, Confederation of Passenger Transport, UK, 1994.
20. DETR (1998) *A New Deal for Transport: Better for Everyone*, HMSO.
21. *Changing Journeys to Work: An Employers Guide to Green Commuter Plans*, Transport, 2000.
22. Goodwin, Cairns, Sloman, Newson, Annable, Kirkbride and Goodwin (2003) *The Influence of Soft Factor Intervention on Travel Demand, Summary Report*, DfT, December.
23. DfT (2004) Guidance and Technical Guidance on Accessibility Planning in Local Transport Plans, London, July 2004.

15
Intelligent Transport Systems

15.1 Introduction

Intelligent Transport Systems (ITS), also known as Transport Telematics, are concerned with the application of electronic information and control in improving transport. We can already see some new systems implemented and can expect the pace of implementation to quicken. With a crystal ball, we can foresee how a typical journey to work may look in 10 years time.

Before leaving home, you check your travel arrangements over the Internet. Often you choose to travel by public transport and you can identify travel times and any interruptions affecting the service. On this occasion, you choose to travel by car as you have an appointment later in the day at one of those old-fashioned business parks that are inaccessible by public transport. There are no incidents recorded on your normal route to work so you do not bother to use your computer route model to select an optimum route for you.

Once in your car, you head for the motorway and select the cruise control, lane support and collision avoidance system allowing you to concentrate on your favourite radio service. Suddenly, this is interrupted by the radio traffic–message channel service giving you information about an incident on your route. You are not surprised when, at the next junction, the roadside variable message sign (VMS) confirms this. Motorway messages really are believable now!

You feel pleased with yourself that you have pre-coded your in-car navigation system with the co-ordinates of your final destination and soon you are obtaining instructions on your best route with information updated from the local travel control centre.

As you near your place of work, you are aware of roadside messages informing you of the next Park and Ride service. You choose to ignore these as you will need to make a quick getaway for your appointment and then check that your travel card is clearly displayed inside the car. You do not want to be fined for not having a positive credit for the city's road pricing and parking service! The same card gives you clearance to your parking space; you activate your parking vision and collision control just to be sure of not scratching the car of the managing director (MD) next to you.

15.2 Using ITS

ITS is a collective name for a number of technology-based approaches that are designed to improve the quality, safety and efficiency of transport networks. One way of categorising these approaches is into the following application areas:

- Traffic management and control
- Tolling

- Road pricing
- Road safety and law enforcement (described in Chapters 11 and 16)
- Public transport travel information and ticketing
- Driver information and guidance
- Freight and fleet management
- Vehicle safety.

All these applications are being developed with assistance from research and pilot implementation programmes in Europe, USA and Japan.

In the UK, in supporting the 10-year plan, the Department for Transport has sponsored a number of intiatives. These include:

- Urban Traffic Management and Control (UTMC)[1]
- Transport Direct
- Travel Information Highway
- Clear Zones
- Smart Cards
- Road User Charging
- Active Traffic Management (ATM)
- ITS Assist.

The UTMC programme is developing an open system design specification for traffic management applications in urban areas. The new specification will enable greater flexibility for procurement and development of new applications.

Transport Direct is a national travel information service to enable people to plan journeys and to compare routes and prices. It covers UK travel by air, rail, coach, bus and car.

The Travel Information Highway will allow local authorities and travel information suppliers to connect to a network and automatically transfer information to other organisations.

The Clear Zones project aims to reduce pollution and traffic in towns and enhance manufacturing and export opportunities by developing innovative technologies and transport solutions.

The Department is supporting the development of multi-function smart cards through the Pathfinder Project, ITSO and the Transport Card Forum.

As well as progressing a charging system for heavy goods vehicles in the UK for introduction in 2007, the Department[12] is also studying the possible introduction of road user charging for all motorised vehicles beyond 2010.

A demonstration project covering the M42 is investigating the introduction of a number of motorway ATM measures including ramp metering (first trialled on the M3 and M27) and peak-time hard shoulder traffic operation.

The Department has listed a number of the benefits of ITS in a series of Traffic Advisory Leaflets.[5] The Department's 'ITS Assist' project provides advice, guidance and information to local authorities on the development and deployment of ITS solutions.

15.3   Traffic management and control

Any traffic management and control needs information on traffic flows, speeds, queues, incidents (e.g. accidents, vehicle breakdowns, obstructions) air quality and vehicle types, lengths and weights. This information will be collected using infra-red, radio, loop, radar, microwave or vision detectors or through the use of probe vehicles. In addition, public and private organisations will provide information on planned events (roadworks, leisure events, exhibitions).

The use to which this information is put depends on the objectives set for management and control. Network management objectives set for urban areas[1] include:

- influencing traveller behaviour, in particular modal choice, route choice and the time at which journeys are made;
- making travel more efficient (safer, less polluting, cheaper and better informed);
- Helping drivers find the best route to their destination including providing information on where to park;
- reducing the impact of traffic on air quality;
- improving priority for buses and light rapid transit (LRT) vehicles;
- providing better and safer facilities for pedestrians, cyclists and other vulnerable road users;
- restraining traffic in sensitive areas;
- managing demand and congestion more efficiently.

The software systems used will include control applications such as SCOOT, SCATS, SPOT and MOTION. The SCOOT traffic adaptive control system is credited with achieving reductions in traffic delay of between 10% and 40% compared with the previous fixed-time control system. These are responsive systems which control a network of traffic signals to meet these objectives. Automatic vehicle location and identification will provide information for giving priority or allowing access to certain vehicles only.

Bus priority relies on vehicle location equipment and a means of communication to traffic control equipment. Vehicle location is achieved using detector loops, roadside beacons, vehicle profile recognition (using inductive loop detection) and global positioning systems (GPS). Communication from loop or roadside beacon is normally wireline based while GPS usually communicate by wireless signal to the traffic signal controller. Enforcement of bus lanes[6,7] is required to ensure bus priority measures are successful.

Air quality monitoring can be linked to a control centre to provide information for decisions on changes to traffic control.[2]

Automated access control is now commonplace on UK urban streets to prevent access for general traffic at certain times of day into sensitive areas. Often control is by lifting barrier or rising bollard using transponders and roadside card readers.

Inter-urban network management systems will have similar objectives but will make greater use of ATM measures. These measures extend variable speed limits, introduced in the controlled motorway system on the M25 between Junctions 10 and 16, to include incident management, maintenance management and access control such as ramp metering. They will include reducing congestion through more efficient use of roadspace such as hard shoulder operation in peak periods or during incidents.

ATM includes motorway incident detection and automatic signalling (MIDAS) with automatic queue detection, a network of CCTV cameras, gantry-mounted Advanced Motorway Indicators. Digital enforcement equipment will open and close lanes, control speeds and provide enhanced driver information.

Regional traffic control centres (RTCC) advise motorists of incidents and alternative routes by VMS, and by radio data system and traffic message channel (RDS–TMC), a signal frequency modulation (FM) radio service broadcasting localised traffic messages and advice to drivers. Their role now is expanding to include controlled motorways and other ATM measures.

In the UK there has been a series of UTMC demonstrator projects in the four towns/cities of York, Reading, Preston and Stratford.

Examples of integrated inter-urban and urban management systems are provided by Munich's COMFORT,[3] Southampton's ROMANSE project and by MATTISSE and the Midlands Driver Information System.[4,9]

MATTISSE is now a partnership between eight local authorities in the West Midlands, Marconi and Mott MacDonald. It has migrated from a service based on user terminals into an Internet-based service through the MATTISSE Viewer which is UTMC compliant and utilises the Travel Information Highway. It provides real-time traffic and travel information on both highway and public transport networks across the region.

Another example is the London Traffic Control Centre Initiative. This includes an extensive network of CCTV and Automatic Numberplate Recognition (ANPR) cameras. Coupled with urban traffic control (UTC) information, the COMET information system uses this network to provide a range of real-time information displayed on a map-based user interface.

15.4   Tolling

Inter-urban motorway tolling provides another approach to meeting network management objectives while obtaining additional revenue that can be invested in transport. The TOLLSTAR electronic toll collection and ADEPT automatic debiting smart cards were early examples of applications.

Currently most of these systems rely on Dedicated Short-Range Communication (DSRC) between roadside equipment and an in-vehicle tag (transponder or smart card). Enforcement is achieved with detection of vehicle licence plates; that is, ANPR using the image processing technology. But the European Commission is arguing that use of satellite technology is the future and has stipulated that by 2008 new systems should be based on satellite technology.

Toll operators want to avoid proprietary solutions that lock them into one manufacturer. Governments also want interoperability between operational systems across national boundaries. The solution is to use official standards supported by many suppliers and this is being achieved in Europe by the adopting of Comité Européen de Normalisation (CEN) standards. Systems formed from more than one supplier are now in usage in many countries.

Interoperability between existing systems can also be achieved by having roadside equipment set up to communicate with several different vehicle tags or the tags themselves are loaded with more than one national application. There are matching standards for Electronic Fee Collection.

The tolling system includes the informing of drivers in advance of the toll collection that a toll is to be paid. The information advises how much to pay and which toll lane can be used. Toll collection is speeded by drivers using smart cards. The efficiency of the toll area is maintained through VMSs at roadside and over lanes and light emitting diode (LED) variable road studs.

A good example of automated toll collection is in Norway which introduced cordon charging systems in Oslo (1990), Trondheim (1991) and Stavanger (2001).[13] Using DSRC, over a million AutoPass electronic transponders (tags) are now in use on more than 200 traffic lanes.

15.5   Road pricing

This includes both national and urban road pricing. The technology used varies but paper based, DSRC and GPS / Mobile Positioning System (MPS) are all in use. The Department's DIRECTS equipment trial will provide a means of testing different technologies and deriving the specification for procurement of charging systems.

The UK Government is considering a form of national pricing on all or a selection of roads for all motorised users. This follows the introduction of national charging systems in several European countries for heavy goods vehicles.

Singapore's electronic zone pricing is an example of an urban system in use for several users having graduated from a paper-based system.

London's Congestion Charging System, implemented in February 2003, uses pre-payment made by both electronic and cross-the-counter methods. Enforcement[10] is achieved using cameras at some 170 entry/exit points around the Central London control zone backed up by Automatic Numberplate Recording. Statistics in early 2004 showed traffic congestion in the control zone down by 40% during charging hours (7 a.m. to 6.30 p.m. weekdays) and the number of vehicles within the zone down by 16%.

15.6 Public transport travel information and ticketing

Travel information is needed by passengers at home or office and also during their journey. London Transport's ROUTES computer-based service offered routeing, timetable and fares information on all public transport services in London through public inquiry terminals and this service is now offered by Transport for London over the Internet.

Real-time travel information is provided in London by the COUNTDOWN system using roadside radio beacons and expanded to cover some 4000 bus stops. Similar systems are now operated in most major towns and cities but use GPS.

ROMANSE includes TRIPlanner interactive enquiry terminals at rail stations in Hampshire with touch screens providing travel information by both train and bus.

Smart card ticketing, such as the Oyster[11] card in London, are now a feature in many public transport operations. Early in 2004, Oyster card added 'pay as you go' capability to use of the card for period tickets. In principle, Smartcard stored-value tickets can also provide a single ticket for car parking and all legs of a journey served by different public transport operators.

PLUSBUS is the first national integrated bus/rail ticket. Prior to PLUSBUS there had been few applications to provide tickets for through journeys. The plan is to progressively roll out the system across the UK. PLUSBUS includes detailed information displays about bus services to provide route information for travellers.

15.7 Driver information and guidance

Driver information systems include the RDS–TMC, initially trialled between London and Paris in the PLEIADES project and elsewhere in Europe in similar European Community-funded projects. There is also the Trafficmaster service which uses infra-red monitors to identify congestion and an in-car visual map-based screen to inform drivers of congestion.

The Scottish Executive's National Driver Information and Control (NADICS) provides an Internet-based Automated Diary Facility (ADF). This includes planned roadworks traffic management and lane closures, occupations and other events on the road network.

The technologies used for transferring travel information include HyperText Markup Language (HTML) for web sites, SNMP for remote monitoring and CORBA for database applications.

Driver guidance systems can inform drivers of their route options and give guidance on navigation. Communication between control centre and the vehicle can be by roadside beacon or by digital cellular radio networks based on Global System of Mobile communications (GSM) as trialled in SOCRATES. Commercial products include Daimler Benz's co-pilot dynamic route guidance system and Philip's Car Systems CARiN. Similar products, the VICS advanced mobile information service, are commonly available in Japan.

Cooperative Vehicle Highway Systems (CVHS) and Advanced Driver Assistance Systems (ADAS) take this a step further. The vision is to provide road–vehicle and vehicle–vehicle communication enabling vehicles to exchange position, speed and direction information and to react as necessary. 'X-by-wire' technologies are enabling steering, throttle and braking systems no longer to rely on mechanical coupling.

This should lead to less stressful driving and to safer and more efficient roads with lane control and close following of vehicles as well assessment of driver performance. The Department of Transport's Road Traffic Advisor project has tested a number of CVHS elements on the M4 and M25.

15.8   Freight and fleet management

Applications for freight and fleet management are based on automatic vehicle location using GPS. Both vehicles and loads can be tracked for security and for optimising fleet usage.

If destinations are known in advance, then the operator can schedule routes and time deliveries more efficiently. The operator can also identify opportunities for consolidating loads or obtaining return loads.

15.9   Vehicle safety

Many vehicle manufacturers or systems suppliers (Ford, Daimler Benz, Lucas, Jaguar) have developed vehicle safety applications.

Autonomous Intelligent Cruise Control uses microwave radar sensors to ensure a safe time interval between your car and the car in front.

Anti-lock Braking Systems are now enhanced with electronic traction control to prevent wheel spin.

Lane Support Systems use radar or video cameras and image processors to detect kerb and lane markings and then warn the driver if he is about to stray from the lane.

Collision Warning and Avoidance Systems detect obstacles, warn the driver and can take action to avoid collision. Toyota uses a milli-wave radar sensor located behind the grille to detect and warn of a crash. Mercedes is offering a 'pre-safe' collision monitoring system.

Driver Monitoring monitors the way in which a driver is controlling the vehicle and provide a warning if performance indicates drowsiness or loss of attention. They also provide a monitor of driver performance and can indicate where further training or attention to risk management is required.

15.10   System integration

Many of these applications can benefit from common standards and a convergence of system architectures. This is because many of these applications use the same technologies and need to pass messages between detectors, control centres, vehicles and roadside information and control systems or between applications.

Data exchange is a key area and the DATEX standard provided by the EC covers this.

In the UK, a planned motorway architecture is provided by NMCS2 and a planned urban traffic management architecture should emerge. Private-sector systems providers will have a strong role in setting future standards and architectures.

15.11   International comparisons

Dr John Miles[8] has researched ITS developments in several countries including the UK, mainland Europe, USA and Japan.

In the UK, the DOT is promoting an open modular architecture for UTMC. The full list of functions that this will support is shown in Figure 15.1. The Highways Agency is aiming to obtain

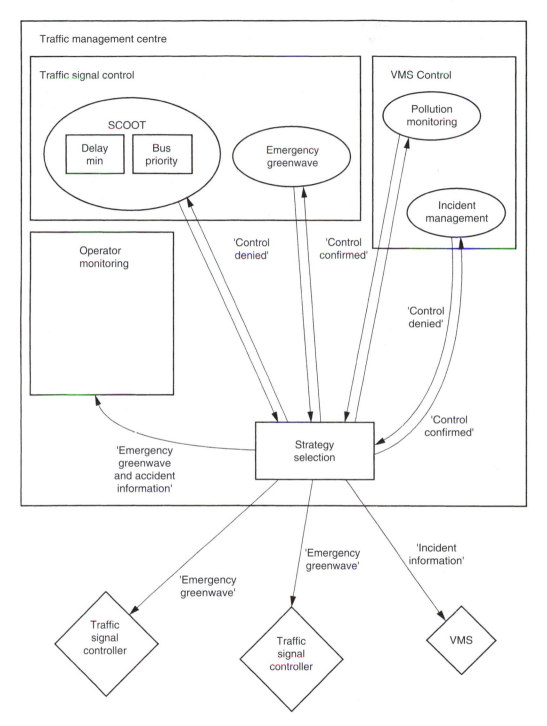

Fig. 15.1 UTMC strategy and selection function.

private-sector involvement in RTCCs. These Centres will control and manage trunk roads and provide information to drivers on roadworks, congestion and incidents.

In Germany, both Munich through the COMFORT project and Stuttgart through the STORM project have developed ITS to an advanced state. The system coverage includes:

- integrated traffic and travel data centre,
- public transport information and management,
- park and ride information,
- UTC and regional traffic control,
- beacon and RDS–TMC traffic information services,
- freight management.

In Turin, Italy, a complete Integrated Road Transport Environmental (IRTE) has been set up to connect 10 traffic and transport control centres to a backbone data communications network. The centres have connections to a 'Town Supervisor' with a common database and a set of forecasting models to improve performance.

The USA has defined an Intelligent Transportation Infrastructure (ITI) and deployment has occurred in four cities: Phoenix, San Antonio, Seattle and New York. ITI covers nine core technologies shown in Table 15.1.

In Japan, the National Police Agency is developing a Universal Traffic Management System (UTMS). The system uses an infrared detector which is used for both control and for two-way

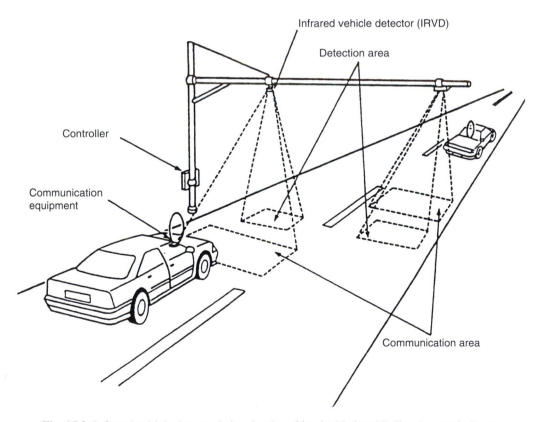

Fig. 15.2 Infrared vehicle detector being developed by the National Police Agency in Japan.

Table 15.1 ITS technologies for metropolitan areas

Regional multi-modal traveller information centres
Traffic signal control systems
Freeway management systems
Transit management systems
Incident management programmes
Electronic fare payment systems
Electronic toll collection systems
Highway-rail crossing protection
Emergency management services

communications as in Figure 15.2. The downlink to the vehicle is capable of carrying traffic information for the driver while the up-link carries a vehicle identifier permitting journey-time monitoring.

References

1. Routledge, Kemp and Radia (1996) UTMC: The Way Forward for Urban Traffic Control, *Traffic Engineering and Control*, London.
2. Hewitt, R (1996) ELGAR, *IEE Conference*, London.
3. Hoops, Csallner and Busch (1996) Systems Architecture for Munich COMFORT, *Traffic Technology International*.
4. Welsh, P and Carden, P (1997) MATTISSE and the Midlands Driver Information System, *TRAFFEX Conference*, PTRC, London.
5. Department for Transport (2003) Traffic Advisory Leaflets ITS1/03–6/03 inclusive.
6. Hewitt, R, Slinn, M and Eastman, C (1996) Using Cameras to Deter the Illegal Use of Bus Lanes in Birmingham, *IEE Colloquium on Camera Enforcement of Traffic Regulations*, London.
7. Turner, D and Monger, P (1996) The Bus Lane Enforcement Cameras Project, *IEE Colloquium on Camera Enforcement of Traffic Regulations*, London.
8. Miles J (1997) UTC Meets ITS; what does the future hold for Urban Traffic Control Systems?, *TRAFFEX Conference*, PTRC, London.
9. George, Glen and Eccleston (2003) The Informed Traveller in the Midlands, *International Conference on Intelligent Transport Systems*, ITS, UK
10. Landles, Wynne and Bozzo (2003) Managing Traffic Through the London Control Centre Initiative, *International Conference on Intelligent Transport Systems*, ITS, UK
11. *The Oyster Smartcard Is a Pearl of a Solution*. Traffic Engineering and Control, January 2004.
12. *Transport 2010, the 10-Year Plan*. the Department of Transport Local Government and the Regions, July 2000.
13. *Fully Automatic Free-Flow Tolling in Norway*. ITS International, November/December 2003.

16

Enforcement

16.1 Introduction

This chapter deals with the increasing role of the traffic engineer in detecting and penalising drivers' contraventions of motoring law. In the past enforcement was, more or less, the exclusive preserve of the police service using the criminal law; however, as traffic levels and the number of offences have increased and police resources and priorities have changed, it has been recognised that the police service has become unable, or unwilling to devote sufficient resources to the enforcement of minor traffic offences. Further there has been a recognition that perhaps the structure and cost of the criminal law process was perhaps inappropriate for the types of offences being dealt with. In simple words, the system was costing too much and was not able to cope with the increasing levels of minor offences being committed.

By the early 1980s, a driver probably had about a 1% chance of having a minor offence, such as parking illegally, being detected. However, although such offences were minor themselves, their consequences could be far reaching. For example, in some city centres illegal parking reached a level where the safe flow of traffic could no longer be maintained and as a result the emergency services were finding that they could not reach buildings to deal with fires and injured people. There is no doubt that because of illegal parking people died who might have otherwise been saved.

16.2 Background

In the past, there was a fairly clear line between the responsibilities for providing and maintaining the highways network, in all its complexity, and for ensuring that the users of that network complied with the rules and regulations that controlled the way that they behaved. The traffic engineer, working for the appropriate highways authority, would determine what rules and constraints would be put in place and it then became the responsibility of the local police authority to detect and penalise the traveller who breached the rules that were put in place. Thus the traffic engineer would determine measures such as:

- speed limits;
- parking, waiting and loading restrictions;
- banned turns and movements;
- weight, height and width limits;
- traffic priorities, such as stop and give-way lines;

- the use of traffic signals to give explicit priority to one stream of traffic, including pedestrians and cyclists, over others;
- priorities, such as bus lanes and bus-only movements.

The police were given the responsibility for enforcing these regulations, along with dealing with other traffic-related offences, such as construction and use regulations, governing the physical design and condition of vehicles and the laws relating to drivers, for example the need to have an appropriate driving licence for the vehicle being used, drink driving and so on. Initially, the standard and only means of dealing with such offences was by summons to the local magistrates' court where the offender would have his offence dealt with in a formal hearing, with the police witness giving evidence against them. By the 1960s the number of such minor offences was threatening to swamp the system and so the government introduced the fixed penalty notice (fpn) to deal with lesser offences. The fpn is a ticket where the motorist is offered the opportunity to pay a fixed financial penalty as an alternative to prosecution at a magistrates' court. When first introduced, the fpn was only applied to stationary vehicle offences, that is parking and waiting restrictions, and a new uniformed force of traffic wardens was created to patrol streets and ticket offending vehicles. However, over time, the scope of the fpn has broadened to include other minor motoring offences, and more recently the concept of the endorseable fpn (fpn(e)) has been introduced to deal with more serious motoring offences where a driver is offered an fpn(e) which levies both a fine and allows the driver's licence to be endorsed with penalty points as an alternative to attending court. The fpn(e) is widely used to deal with drivers who are detected using speed and red-light cameras.

Despite the wider use of fixed penalties, it was becoming apparent that, with changing government priorities for policing, the police service was not able to enforce existing regulations, let alone the response to increasing pressures for ever more restrictions to increase parking control and manage traffic movement for safety and amenity reasons. The 1991 Road Traffic Act[1] was the first measure that sought to address the mismatch between the need for enforcement and the supply of police services by allowing local authorities to opt to move contraventions of parking, waiting and loading restrictions from the criminal law to being a civil offence. Since that time more and more minor traffic offences have been 'decriminalised' and the responsibility for their enforcement has passed to the local highway authority.

16.3   Parking, waiting and loading

The Road Traffic Act 1991 has transferred responsibility for dealing with stationary vehicle offences to the local authority. The powers came into effect in London in 1994, with the London Borough of Wandsworth being the first to adopt them in late 1993. The powers are also available to other local authorities in Britain and are gradually being adopted by local authorities outside London, driven in part by decisions by police authorities to withdraw traffic warden services and use the resources for other tasks.

The powers allow parking attendants employed by the local authority to patrol and issue a penalty charge notice (pcn) to vehicles found in contravention of waiting, loading and parking restrictions. The pcn identifies the contravention and levies a penalty charge. If paid quickly the charge is discounted by 50%; however, if the driver does not pay the charge can be subject to a surcharge if paid, or collected late. The system allows a driver to challenge the pcn at two levels. First the driver can formally apply (make representations) to the issuing authority requesting that the charge is cancelled. Strictly speaking those considering representations are only obliged to consider matters of law but most local authorities will also consider extenuating circumstances within reason. If the motorist is not satisfied with the outcome of the representations, they can then appeal to an independent adjudicator for a final decision.

There is no explicit appeal process beyond the adjudicator, although there have been a few applications for a judicial review on a point of law. Experience suggests that about 5% of drivers feel strong enough to take a case to adjudication and the outcomes of hearings are quite evenly balanced between councils and motorists.

The Road Traffic Act 1991 also allows councils to use vehicle removals and wheel clamping as enforcement tools but, outside the metropolitan areas, take up of these powers has been quite limited.

The Road Traffic Act 1991 did not allow local authority attendants to deal with obstructive parking outside regulated streets. This remains the responsibility of the police who can deal with such behaviour as a highway obstruction. The original Act also created a paradox in that only non-endorseable offences had been decriminalised. This meant that, although a parking attendant could issue a pcn for a vehicle parked on a yellow line say, they could take no action against the more serious offence of a vehicle parked within a pedestrian crossing, an offence which would usually attract an fpn(e). The law was subsequently modified to allow such an offence to be enforced by either the parking attendant or the police service.

Within London the use of a pcn has been extended by the London Local Authorities Act 2000[2] to allow the system to be used for illegal parking, where the contravention has been detected using CCTV cameras, for example at a school entrance. A camera is sighted to give surveillance on a length of street where there is a persistent and dangerous parking problem and periodically the image is viewed remotely by a parking attendant who will note any offending vehicles. The images are recorded to provide corroborating evidence, and at the end of the surveillance period the observer enters the observed contraventions into a computer system which will trace the registered vehicle keeper via the Driver and Vehicle Licensing Agency (DVLA) and then issues a pcn by post. The system allows regular but essentially random surveillance of hot spots and because the paperwork is done after the event then greater productivity would be the case if the attendant had been at the side of the road. It could be argued that there is something slightly sneaky about the use of cameras in this way, but if the observation camera is in the open and if one relied only on a staff presence there is little doubt that drivers would take the lack of a visible parking attendant as permission to park.

16.4   Bus lanes

Camera-based enforcement systems were introduced into London in a co-operative venture between the then Traffic Director and the Metropolitan Police. The systems relied on a combination of fixed and mobile, bus-mounted, cameras which recorded activity on bus lanes for subsequent evaluation. Staff employed by the Traffic Director would review the tapes and determine whether or not they believed that an offence had been committed. If so they would obtain the keeper details from DVLA and pass cases over to the Metropolitan Police who would review the evidence and issue an fpn if they were satisfied that an offence had been detected.

In the London Local Authorities Act 1996 modified the law to allow London Boroughs to deal with offences in bus lanes using a pcn. Using CCTV cameras council staff follow a procedure similar to that outlined above to detect unauthorised vehicles in bus lanes and issue pcns by post to those committing the contravention. These powers were later extended to Transport for London following the establishment of the Greater London Authority.

Evidence from London suggests that this form of enforcement is very effective with non-compliance dropping by about 80% where cameras are used. The Traffic Management Act 2004[3] has created new powers to allow local authorities to enforce bus lane non-compliance and other minor moving offences similar to those created for London by the London Local Authorities and Transport for London Act 2003.

16.5   Other traffic offences

The London Local Authorities and Transport for London Act 2003 has created additional powers to allow London Local Authorities to deal with a range of other minor offences using camera detection and a pcn. The Act also allows council officers to issue fpns for certain other offences. The camera will monitor regulations, such as yellow-box junctions and banned turns, and offending vehicles detected using the same procedures as for the camera detection of parking offences will be sent a pcn by post. These are new powers which have yet to be implemented at the time of writing. It is likely that such the powers have been extended nationwide by the Traffic Management Act 2004.

16.6   Congestion charging

In February 2003 London launched its long-awaited system for charging motorists a surcharge to enter the centre of London using powers in the Greater London Authority Act 1999. Such a proposal had first been researched as far back as 1973 and the area chosen for charging was that which had been identified 30 years ago. However, the mechanisms for charging, detecting offenders and for penalising offenders were radically different from what had been considered 30 years ago. In the 1970s, it had been assumed that those who wished to pay for a 'supplementary licence' would buy a paper permit which would be displayed on their windshield and detected by bands of traffic wardens carrying out spot checks at the side of the road and issuing fpns to offenders.

The system implemented in London is a high-tech system where the payment and enforcement systems are integrated and largely paperless. A driver who wishes to drive in the restricted area can pay the charge via the internet, using a mobile phone, by post, at a street-side pay station or at a retailer. The charge can be pre-paid or paid up to midnight on the day of travel. When the charged is paid the vehicle registration mark is entered into a database which also contains exempt vehicles, such as buses and taxis and emergency service vehicles.

Throughout the charged period traffic entering, leaving and driving within the charged area is monitored by a large number of cameras linked to an automatic numberplate recognition system which identifies and records vehicle registration plates automatically. At the end of the day the list of seen vehicle numberplates is compared with the list of permitted/paid vehicles and any mismatches are identified as potential offenders. The photographic evidence is checked manually and any errors eliminated. The pcns are then sent to offending vehicles. Although the system philosophy is not part of the legislation it would seem logical that if any other city were to introduce congestion charging, they would look for a similar technique.

16.7   Comment

It does seem quite inevitable that as traffic volumes continue to grow, despite the oft-stated public policy of reducing car dependence, so our roads will be subject to more and more controls and restrictions to manage the way precious road space is allocated and used. Equally inevitably this will result in many users committing minor traffic transgressions either through ignorance, confusion (and who has not looked at an unfamiliar traffic sign and not understood its meaning) or selfishness. The police service is unlikely to have either the resources or the inclination to deal with such offences and so inevitably such transgressions go largely unremarked except perhaps by other road users who are inconvenienced or put at risk by the contravention, and so it is likely that there will be an increasing tendency to make the highways authorities who implement the regulations also responsible for ensuring that they are complied with. In the UK, at the present time, only a police officer in uniform has the power to stop a vehicle and so it seems inevitable that local authorities will look to the use of technology to enable them to detect and penalise such contraventions.

In the UK, there does seem to be an attitude to traffic law enforcement which requires that either offences are dealt with by the police or by the civil authorities, meaning that if an offence is 'decriminalised', the police lose their powers to deal with it. This is not the case in other countries where the decision to have municipal enforcement does not automatically result in the loss of police powers, meaning that although the police may no longer pro-actively seek out minor traffic offences they are still able to deal with offenders if necessary. This is particularly pertinent given that only a police officer can actually make an offender stop, that there is quite a lot of evidence which suggests that people who commit more serious offences have a propensity to also ignore minor traffic regulations and the simple fact that something over 2 million vehicles in the UK have incorrect or no keeper details at DVLA.

References

1. UK Government (1991) *Road Traffic Act 1991*, HMSO, London.
2. UK Government (2000) *London Local Authorities Act 2000*, HMSO, London.
3. UK Government (2004) *Traffic Management Act*, HMSO, London.
4. UK Government (2003) *London Local Authorities and Transport for London Act 2004*, HMSO, London.

17

Statutory Requirements

17.1 Introduction

Traffic engineering activity, particularly on the highway, is controlled by an enormous number of laws and regulations which, taken together, set out:

- the rights of landowners;
- the rights and responsibilities of road users;
- the rights and responsibilities of the highways authority and, by inference, the traffic engineer acting on their behalf;
- the powers of the police to enforce the law, particularly as it relates to traffic and travellers.

In the UK, legislation is formulated in a number of ways. Acts of Parliament or primary legislation provide the main source of the laws that govern roads and traffic. Often, however, Acts of Parliament only provide general and non-specific enabling powers. These powers allow something to be done, but without specifying the manner in which that thing should be done. In these circumstances secondary legislation, in the form of a Statutory Instrument, is required to give effect to the primary legislation contained in a Parliamentary Act.

A good example of this two-tier relationship can be found in the *Road Traffic Regulation Act 1984*.[1] This Act gives a highway authority the power to erect traffic signs but makes no provision as to the form or type of signs allowed. These have been subsequently defined in a Statutory Instrument SI 2002/3113.[2]

This two-tier system of legislation avoids detailed technical matters becoming the subject of an Act of Parliament. It also means that the legal requirements can be adapted more quickly in response to changing circumstances and technology, without the need to enact new primary legislation, provided that the underlying need remains the same. Thus, for example, the 2002 Signs Regulations replaced an earlier 1994 Statutory Instrument which in turn replaced a 1981 document, reflecting the change in sign designs over the intervening years, and the introduction of brown tourist signing, but without a change to the underlying requirement for signing.

The government also issues countless circulars, advisory notes, guidelines and technical papers which formally set out guidance on the interpretation of the law and advise on the procedures to be followed in implementing new systems of traffic management and control.

The government also acts to regulate and control highways and traffic on national basis. However, it is also possible for a local authority to seek legislation which has only a local effect.

Thus, for example, it is an offence to park on the footway within the Greater London area, because of a local Act (*The GLC General Powers Act 1969*[3]) and although equivalent powers exist nationally, in the *Road Traffic Act 1974*,[4] this is an enabling power and the government has never brought the powers into effect by enacting an appropriate Statutory Instrument. Local powers can be granted by means of a local government act which applies specifically to an area, or by means of a local by-law.

17.2   The scope of legislation

The legislative process is ongoing and it is inevitable that whatever is written here will be overtaken by events very quickly. Recognising this, we have attempted to set out the main principles of legislation and to identify the key acts and regulations which set out the main body of relevant law. With an extensive body of written law, the tendency in recent years has been to only bring forward primary legislation when it is required to legislate in some new area or to update or remove law which has become obsolete.

The main areas of legislation of relevance to the traffic engineer are:

- the Highways Acts, which mostly deal with the provision and maintenance of highways and roads;
- the Road Traffic Acts, which set out the laws controlling road traffic, often including laws relating to vehicles and drivers;
- the Road Traffic Regulation Acts, which primarily set out the powers that a highway authority has to regulate and control traffic on roads.

17.3   The Highways Acts

The *Highways Act 1980*[5] provides many of the basic laws relating to the provision and maintenance of highways. The words highways, roads and streets are in common usage and are used interchangeably by most people. However, the words have different meanings in law, although many may find the definitions and differences are far from clear.

17.3.1   A HIGHWAY

We might reasonably expect to find a workable definition of a highway in the Highways Acts. The converse is true. The *Highways Act 1980* is far from helpful saying that 'a highway means the whole or part of a highway other than a ferry or waterway'! A more useful definition is offered in '*An Introduction to Highway Law*'[6] which suggests that, based on common law, a highway is, in simple terms, a route which everyone can use whenever they wish, without let or hinderance and without charge.

The only exceptions to the legal requirement that highways should be available for use without charge are those few highways which are subject to special Parliamentary Acts, such as the Dartford Tunnel and Bridge, where the highway was created by a special Act which included the power to charge a toll for use of the facility. Historically, this type of highway, which has usually been an estuarial crossing, has been provided using a special Parliamentary Act. However, the *New Roads and Street Works Act 1991*[7] created a more general power which allows a highway authority to enter into an agreement with a third party, called a concession agreement, to allow the third party to build a new road and then levy a toll for its use. The concessionaire would be free to set the toll at whatever level they wished, except for an estuarial crossing, where the government can set an upper limit on the toll.

17.3.2   A ROAD

The legislation is similarly vague in offering a simple definition of a road. The *Road Traffic Regulation Act 1984* says that a road means 'any highway or of any other road to which the public has access'. Thus a highway is a road and a road is a road but, as far as the law is concerned, a road is only a road if the public has access over it.

This is an important difference. The public have a right to use a highway. However, on a road, which is not a highway, the public may have access but not necessarily the right to pass and repass, without let or hindrance. For example, this definition could apply to a service road on a retail development, where the public are allowed access, but not as a right. Although not offering a very useful definition of what a road is, this definition has the useful effect of allowing the application of road traffic law to roads, such as service roads and private roads which are not highways.

17.3.3   A STREET

The legal definition of the word street given in the *Highways Act 1980* and repeated in the *New Road and Street Works Act 1991* is:

(a)   any highway, road, lane, footway, alley or passage;
(b)   any square or court;
(c)   any land laid out as a way, whether it is for the time being formed as a way or not.

This perhaps could be interpreted as any land where the use of the land is to provide a means to get to adjoining property. Thus streets may not be roads or highways but could give access to property.

17.4   The Road Traffic Acts

The Road Traffic Acts deal with drivers and vehicles. The Acts specify:

- the offences connected with the driving of a vehicle;
- the qualifications required for drivers;
- the construction and use of vehicles.

The latest Road Traffic Act, the *Road Traffic Act 1991*,[8] in addition to creating a number of offences also created a new decriminalised system of parking. The law allows local highway authorities to take over responsibility for the enforcement of most stationary vehicle offences from the police. The Act also moved the offences from the ambit of the criminal law to the civil law, with the penalty dealt with as a debt, rather than as a fine.

The *Road Traffic Act 1991* also created a new administrative system for the main road network in London. The Act allowed the creation of the Priority Route Network with the position of Traffic Director for London to manage the network.

The Act also sets out the powers which provided for the use of red light and speed cameras to record, respectively, drivers who fail to stop at a traffic signal or who are speeding.

17.5   The Road Traffic Regulation Act

The *Road Traffic Regulation Act 1984* provides the legal basis for management of the highway network. The main parts of the Act are as follows:

- Part I: Provides the general powers for the regulation of traffic and gives powers which allow most parking and traffic control functions.

- Part II: Deals with special types of traffic regulation.
- Part III: Deals with pedestrian crossings and playgrounds.
- Part IV: Deals with provision of parking.
- Part V: Provides the powers which allow a highway authority to provide traffic signs.
- Part VI: This contains the powers relating to controlling the speed of traffic.
- Part VII: Although there is a basic right to 'pass and repass' on the highway, the highway authority has powers, under this part of the Act, to erect obstructions, for example bollards, to limit the types of traffic that can use a highway.
- Parts VIII and IX: Deal with enforcement powers.

These powers generally apply to Britain. There are separate powers in Northern Ireland.

17.6   Other legislation

The form and structure of the UK legislative system means that powers which affect the highway can appear in any Parliamentary Act and it would be impossible to authoritatively identify all the relevant statutes in a work of this sort.

There is, however, a body of law dealing with the rights of other bodies, such as the public utilities, to install and maintain their services in the highway. The traffic engineer needs to be aware of the rights and responsibilities of these organisations.

The *New Roads and Street Works Act 1991* deals with two areas of legislation. The Act provides powers for the provision of new roads, financed by the private sector. The Act allows that bodies, other than the highways authority, can provide a new road and recover the cost of that road by charging a toll for its use.

Parts III and IV of the Act are concerned with street works. Prior to the passing of this Act, the many agencies with access to the highway could, and did, carry out their activities in a fairly unco-ordinated way, often with little thought as to how their actions might affect traffic movement. The rationale of the 1991 Act is to provide a basis for regulating and managing the work of highways authorities and statutory undertakers, so that, as far as is practical, activities can be co-ordinated so as to minimise the adverse impact on travel.

References

1. *Road Traffic Regulation Act 1984*, HMSO, London.
2. UK Government (2002) *Traffic Signs Regulations and General Directions 2002* (Statutory Instrument 2002/3113), HMSO, London.
3. *Greater London Council General Powers Act 1969*, HMSO, London.
4. *Road Traffic Act 1974*, HMSO, London.
5. *Highways Act 1980*, HMSO, London.
6. Orlik, M (1993) *An Introduction to Highway Law*, Shaw and Son.
7. *New Roads and Street Works Act 1991*, HMSO, London.
8. *Road Traffic Act 1991*, HMSO, London.

Index

AADT 46
Abnormal load 81, 82
Access control 211
Accessibility Audit 188, 193, 194
Accessibility Planning 188, 193
Accessibility profile 183
Accessibility Ratio 197
ACCESSION 188, 193
Accident costs 147
Accident records 126
Accident reporting 126
Accident savings 144
Accident severity 123, 133, 148, 153
Accidents – contributory factors 126, 130
Accidents – data analysis 131
Accidents – data presentation 139
Accidents – factors 123
Accidents – histograms 133
Accidents – severity 123
Accidents – speed reduction 153
Active bus priority 173
Active traffic management 210
Adaptive, Traffic Responsive UTC Systems 73,
    92, 93
Advanced Driver Assistance System 213
AIMSUM 46
Air quality monitoring 211
Alignment 49, 50, 78, 79, 85–87
Annual average daily traffic, see also AADT 46
ANPR 26
ARCADY (Assessment of Roundabout Capacity
    And DelaY) 86, 87, 89, 103, 179
Area action plans 139
Area of outstanding natural beauty (AONB) 80
Assignment 38
ASTRID (Automatic SCOOT Traffic Information
    Database) 101
ATCO CIF 193
Automatic Numberplate Recognition 212
Automatic traffic counter 76
Automatic Traffic Counts 10

Boots 205
BS 6571 part 1 115
BS 6571 part 2 115
BS 6571 part 3 118
BS 6571 part 7 119
BS 6571 part 8 114
Bus Lane enforcement 220
Bus patronage 202
Bus priority – vehicle location 211
BUS TRANSYT 173
Bus, 12-metre European Low Floor Bus 82,
    175
Bus, Advance Areas 175
Bus, Boarders 175, 176
Bus, Bus-only streets 167, 169, 172
Bus, Busways 167, 169, 172, 178
Bus, gates 172, 178
Bus, information 168
Bus, lanes 68, 69, 167, 169, 173, 175, 178, 211,
    219, 221
Bus, priority 69, 92, 98, 99, 101, 103, 166, 167,
    173, 175, 178, 179
Bus, regularity 168
Bus, shelters 177
Bus, stop clearways 175
Bus, stop 69, 98, 167, 169, 172, 173, 175–179

Cableless linking 90
Cant 76, 78
Capacity restraint assignment 44
Capacity 49, 52–58, 66–70, 73, 82, 84, 86,
    87–90, 92, 101, 103, 169, 173, 175, 192
Car clubs 204
Casualties 123
Casualties – child pedestrians 138
CCTV (Closed Circuit TeleVision) 75
Centrifugal 77
Channelisation 72, 85
Charged for parking 115
Chicanes 160
Chi-squared test 140

Circular 7/75 63
Classified counts 12
Clear Zones 210
Clockwork parking meter 115
Collision Warning 214
Conflict crossing 58, 83
Conflict diagram 82, 83
Conflict diverge 54, 83
Conflict merge 54, 55, 58
Conflict studies 140
Congestion charging 68, 69, 75, 201, 221
Consultation 156, 179
Continuous observation parking surveys 35
Contraflow bus lanes 169, 171, 178
Control test 149
Controlled parking zone 110
Cooperative Vehicle Highway System 213
Crossfalls 76, 78, 87
Cross-road layouts 84
Cruise speed 92, 103
Cycle advance area 88, 98
Cycle time 90, 98, 99, 101, 103
Cyclists 49, 69, 73, 88, 166, 167, 169, 172, 178

Daily flow profile 67, 68
Data patching 21
DATEX 214
Deflection 50
Degree of saturation 93, 103
Demand management 66–8, 201
Demand, flow 52, 103
Departure 78, 101
Design, flows 50–52, 55, 82
Design, speed 50, 57, 76, 77, 79, 81, 84, 85
Design, vehicle 81, 82
Design, year 52
Detectors 10, 210
Deterministic Theory 52, 53
Development Plans 181
Direction signs 61
Disability Discrimination Act 1995 112
Disabled driver parking 112
Disc Parking 114
Diverge 54, 56, 58, 83
Diversion curve 43
DMRB 49, 58, 76, 77, 84, 86, 89
Dot-matrix signs 75

Driver Information 213
DYNASIM 46

Effective green time 102, 103
Electronic parking meter 117
EMME/2 45
Enforcement 166, 176
Enforcement – traffic speeds 157
Engineering measures 69
Environmental assessment 197
Environmental impacts 193
Environmental, Traffic Management Schemes (ETMS) 66
Equilibrium approach 45

False channels 79
Fault monitoring 101
First year rate of return 147
Fixed penalty notice 219
Fixed-time 73, 90–3, 101, 103, 173
Floating car surveys 24
Flow-capacity relationship 52
Free parking 114
Freight management 214
Frontagers 166, 178
Full overtaking sight distance (FOSD) 77, 80
Fully adaptive systems 92, 93
Furness 41

Gateways 161
Generalised cost 44
Geographic information system 185
Ghost island 55, 72, 73
Grade separation 70, 72–74, 82
Gradient 50, 79, 102
Gravity model 44
Green time 73, 90, 98, 101–103, 169
Green wave 73, 101
Growth factor 40
Guided bus 167, 172

Headways 50, 167, 168, 179
Highway 224
Highway Code 178, 179
Highway link design 76, 89
Home Zones 70, 161
Horizontal Alignment 49, 76–78, 80
Horizontal deflection 50, 160
Hurry call 90, 101

IHT – accident investigation 142, 163
IHT – urban design 163
Induced traffic 200
Inductive loop 10, 90, 93
Information signs 61
Instation transmission unit (ITU) 93
Intersections: general 76, 80, 84, 89
Isochrones 185
ITS Assist 210

Journey speed 24
Junction, capacity 54, 55, 57, 86, 90
Junction, delay 24, 86, 89, 90, 93, 101–104
Junction, flow diagrams 84
Junction, types 54, 70, 71, 82

K' value 77, 79
Keeping Buses Moving 169, 179

Level of service 49, 50
Link profile unit (LPU) 93, 99, 101
LINSIG 103, 104, 179
Local accessibility 196
Local Development Documents 182
Local Plans 181
Location sampling 141
London Local Authorities Act 2000 220
London Traffic Control Centre 212
Loop detector 90, 93, 94,96, 98–100
Lorry control 162
Lost time 102

Manual classified counts 14
Mass action plans 139
Matrix estimation 45
MATTISSE 212
Merge 54-56, 58
Micro simulation 68
Micro-processor 90, 92
Microsimulation 45
MIDAS 211
Minerals Planning 180
Mini-roundabout 72, 73, 82, 86, 87, 89
Mobile phone parking 121
Modal share 38
Modal split 38
MOLASSES 150
MOTION 211

MOVA 73
M-parking 121

National Road Traffic Forecasts 40
Network accessibility 196
Network management 211
Noise 198
Numberplate surveys 14

Off-set optimiser 98
Off-side priority 72, 86
On-street parking 105, 120
Operational performance 192
Origin–destination matrix 20, 41
Origin–destination surveys 17
OSCADY (Optimisation of Signal Capacity And DelaY) 103, 104, 179
Outstation Transmission Unit (OTU) 93, 99
Overlap phases 173, 174
Overtaking sight distances 77, 80

PARAMICS 46
Park and ride 166
Parking beat surveys 30
Parking charges 203
Parking control systems 113
Parking meter 115
Parking occupancy surveys 29
Parking supply surveys 29
Parking surveys timing accuracy 34
Parking voucher 119
Passive bus priority 173
Pay and display machine 118
Pay-on-foot 121
PCU (Passenger Car Units) 49, 50, 51, 102
PCU/vehicle ratios 102
Peak period analysis 47
Peak spreading 67
Pedestrian crossing 49, 66, 69, 70, 74, 76, 85–88, 90, 92, 98, 99, 178, 179
Pedestrian flows 87
Pedestrian subways 88
Pedestrians 49, 66, 69, 73, 85, 86, 88, 89, 166, 172, 178
Penalty charge notice 219
People with disabilities 67, 69, 166, 178
Performance index (PI) 103
Performance indicators 204

Personal injury 123
Personal travel planning 204
Pfizer 205
PIA 123
PICADY (Priority Intersection Capacity And DelaY) 53, 58, 86, 103
Plan selection 93
Planning and Compulsory Purchase Bill 182
Planning and Policy Statements 180
Planning Policy Guidance 180
PLUSBUS 213
Poisson test 140
Police authorities 124
Potential accident reduction 139
Pre-signals 175, 178
Pressure tube 10
Primary legislation 223
Priority give-way 58, 84
Priority junctions 58, 70, 72, 73, 82, 84, 86, 89
Private non-residential (PNR) 68
Probe vehicles 26
PTAL 196

Queue length 24, 52, 53, 99, 173, 179
Queue Relocation 175

Radar speed meter 76
RDS–TMC 211
Real-time Information 213
Recording numberplates 37
Regional Policy Guidance 180
Regional Spatial Strategies 180
Regional traffic control centres 211
Regional Transport Strategy 182
Regression to the mean 131
Regulatory signs 61
Reliable track 167
RFC (Ratio of Flow to Capacity) 52, 53
Ring junction 87
Road 225
Road accesses 185
Road accidents 123
Road hierarchy 185
Road humps 157
Road markings 63, 72–75
Road pricing 201, 203
Road safety audits 148

Road safety plans 148
Road safety targets 134, 137
Road safety – effect of speed 148
Road Traffic Act 1991 219
Road Traffic Regulation Act 1984 108
Roads – inter urban 47
Roads – main urban 47
Roads – recreational 47
Roadside beacons 211
Roadside interview surveys 17
Roadworks 210
Rotating board sign 75
Roundabouts 58, 69–73, 86, 87, 103, 169
Route action plans 139
Rumble devices 157

SACTRA 200
Safe Routes to School 205
Safer Parking scheme 111
Sampling errors 47
Saturation flow 50, 51, 58, 93, 99, 101–104
Saturation Occupancy (SatOcc) 93, 99
SATURN 45
SCATS (Sydney Co-ordinated Adaptive Traffic System) 93, 211
School travel plans 204, 207
SCOOT (Split Cycle time and Offset Optimisation Technique) 45, 73, 75, 90, 92–99, 101, 103, 173, 175, 211
Scoping Study 188
Seasonal variation 47
Selective vehicle detection (SVD) 88, 90, 98, 101, 173, 177
Service bays and lorry parking 112
Service flow rate 50
Set back 169, 173, 175
Shared surfaces 164
Shopping model 192
Sight distances 77, 80, 84, 85
Signalled junctions 82, 83, 87, 90, 98, 101, 103, 104, 173
Single sites 139
Sites of special scientific interest (SSSI) 80
Smart Cards 210
Social inclusion 204
Soft measures 204
Southampton University Hospital 205
Space mean speed 23

Speed cameras 75
Speed cushions 157
Speed humps 157
Speed limit 148
Speed reduction 86, 87, 163
Speed surveys 22, 76
Speed tables 157
Speed, 85th percentile speed 76
Split optimiser 93
Spot speed 22
SPOT 211
Stage 73, 90, 92, 93, 98, 99, 103, 174, 178
STATS 19 124, 126
Statutory Instrument 223
Steady state theory 52
Stick diagrams 131, 132
Stochastic assignment 44
Stockley Park 205
Stopping sight distance (SSD) 77, 80
Street 225
Structure Plans 181
Superelevation 76-79, 87
Sustainable development 199
SVD 88, 90, 98, 101, 173, 177
SVD Detector Location 98
Swept path 82, 84, 87-89, 178

T junction 82-84
Tactile surfaces 89
Temporary signs 64
TEMPRO 41
The highway code 59
Time and distance diagram 91, 92
Time mean speed 22
Time-dependent queuing theory 53
Tolling 212
Town and Country Planning Act 180
TRACK 82, 89
Traffic 1
Traffic adaptive control 211
Traffic calming 50, 67, 78, 167
Traffic calming – overseas 154
Traffic calming – ranking 154
Traffic calming – signing 157
Traffic calming – site selection 154
Traffic calming – UK 154
Traffic counts 10
Traffic engineering 2

Traffic Impact Assessment 185
Traffic intensity 52, 54
Traffic Management Act 2004 220
Traffic metering 73, 175
Traffic model 43
Traffic model *calibration* 45
Traffic Order Procedure regulations 110
Traffic Regulation Order 113
Traffic signals 50, 58, 69–71, 73, 90, 92, 98,
    101–104, 167
Traffic signs 73, 75
Traffic signs regulations and general directions
    2002 59
Transformed curve 52, 53
Transition curve 78, 79
Transponder 90, 100, 101
Transport assessment 185
Transport direct 210
Transport impact 192
TRANSYT (TRAffic Network StudY Tool) 92,
    93, 103, 173, 179
Travel awareness 204
Travel information 213
Travel Information Highway 210
Travel Plans 204, 205
Travel Plans – examples 205
Travel Plans – impact 207
Travel Plans – preparation 206
Travelwise 206
TRAVL 190
TRICS 190
Trip assignment 38
Trip distribution 38
Trip generation rates 191
Trip generation 38
Trip-end model 43
TRIPS 45
Turning circle 81, 82, 88
Turning movements 14

Uncontrolled non-priority junctions 70, 72
Under-counting 34
Underground services 81
Unitary Development Plans 182
Urban design 163
Urban regeneration 203
Urban safety management 163
Urban Traffic Management and Control 210

Use Classes 188
UTC (Urban Traffic Control) 73, 75, 90, 92, 93, 98–99, 101, 103

Vehicle actuation 73, 90
Vehicle categories 13
Vehicle detectors 73, 88, 90, 92, 93, 103
Vehicle removal 220
Vehicle safety 214
Vehicle speed 50, 67, 70, 76
Vehicle turning circles 81, 88
Vertical alignment 49, 76, 79, 87
Vertical curvature 77, 79
Vertical deflection 157
Video survey 26
Village speed control 163
Virtual bus lane 169
Visibility envelope 84
Visibility requirements – road accesses 185, 187

Visibility splay 84, 85
VISSIM 46
VMS (Variable Message Sign) 75, 101, 212

Wardrop
Warning signs 60
Waste and Minerals Local Plans 182
Weaving section 58
Wheel clamping 220
Whole life care of car parks 111
With-flow bus lanes 167, 170
Workplace travel plans 204, 207

'x' and 'y' distances 84

Zones – 20 mph 161
Zones – home 161
Zones – traffic surveys 20